AI Applications to Power Systems

AI Applications to Power Systems

Editor

Tek Tjing Lie

MDPI • Basel • Beijing • Wuhan • Barcelona • Belgrade • Manchester • Tokyo • Cluj • Tianjin

Editor
Tek Tjing Lie
Auckland University of Technology
New Zealand

Editorial Office
MDPI
St. Alban-Anlage 66
4052 Basel, Switzerland

This is a reprint of articles from the Special Issue published online in the open access journal *Energies* (ISSN 1996-1073) (available at: https://www.mdpi.com/journal/energies/special_issues/AI_applications_power_systems).

For citation purposes, cite each article independently as indicated on the article page online and as indicated below:

LastName, A.A.; LastName, B.B.; LastName, C.C. Article Title. *Journal Name* **Year**, *Volume Number*, Page Range.

ISBN 978-3-0365-3653-8 (Hbk)
ISBN 978-3-0365-3654-5 (PDF)

Contents

About the Editor

Tek Tjing Lie received his Bachelor's degree in Electrical Engineering from Oklahoma State University, USA, in 1986. He also received his Master's of Science and Ph.D. degrees in Electrical Engineering from Michigan State University, USA, in 1988 and 1992, respectively. Professor Lie is the Deputy Head of School of Engineering, Computer and Mathematical Sciences, Auckland University of Technology, New Zealand. His research interests include power system control and operation, AI application to power systems, deregulated power systems, smart grids, and renewable energy systems. He has authored/co-authored more than 250 journal and international conference papers. Prof Lie serves as Guest Editor of Energies and Electric Power Systems Research Journals. He also serves as Associate Editor of Modern Power System and Clean Energy and Energies journals, respectively. He is the Chair of the IEEE New Zealand North Section and serves as an organising committee member of several international conferences.

Preface to "AI Applications to Power Systems"

The energy system of the future is a work in progress. Many countries around the world have made a target for their energy system to be completely renewable. How the power system eventually forms, such as the business models, the key players, and its architecture, as well as how it works, will be dependent on the outcomes of trends, forces, regulations, and strategic actions by many diverse players in the energy sector.

The digitalisation of the traditional generation plan and transmission has been proposed. However, there is no existing work in the literature on the digitalisation of wind turbine generation or solar photovoltaic farms. Future power systems will be very complex due to the multiple forces affecting various levels of the system, especially the distribution network level. The systems are composed of thousands of local distribution areas operated by distribution operators (suppliers) on top of the consumers, etc. Thus, although it may not appear too different, the power flows are no longer just one way, going from the bulk power system to the consumer end. In fact, the power flow will be very hard to trace because it can be from one consumer to another, the consumer to the grid, or the grid to the consumer, and some will be mobile/random due to the charging/discharging of electric vehicles. These types of renewable energy resources (solar PV and wind turbine generation) are incredibly dependent on nature (wind speed, wind direction, temperature, solar irradiation, humidity, etc.). Thus, the outputs are highly stochastic in nature. Data science techniques for handling real-time big data will ideally fit this stream. Furthermore, integrated systems modelling methods and concepts are needed for the study of the self-organisation, complexity, emergent properties, and dynamical behaviour of complex systems for their holistic understanding, management, and development based primarily on neural networks, fuzzy and soft systems/fuzzy cognitive maps, network modelling, and mathematics. Other advanced applications in computational early detection of mastitis and computer-based decision support systems for complex systems are also needed. Due to the scale of the network and the amount of data that needs to be digitised, new techniques in data mining and AI approaches are needed in order to analyse and predict the behaviour of these complex systems.

I would like to thank the staff and reviewers for their efforts and input. The task of editing and selecting papers for this collection was found to be both stimulating and rewarding [1,2,3,4,5,6].

Tek Tjing Lie
Editor

MDPI

Editorial

Editorial to the Special Issue "AI Applications to Power Systems"

Tek-Tjing Lie

Department of Electrical and Electronic Engineering, Auckland University of Technology, Auckland 1010, New Zealand; tek.lie@aut.ac.nz

Abstract: This Special Issue consists of the successful invited submissions to *Energies* on the very topical subject area of "AI applications to power systems".

Citation: Lie, T.-T. Editorial to the Special Issue "AI Applications to Power Systems". *Energies* **2021**, *14*, 5667. https://doi.org/10.3390/en14185667

Received: 13 August 2021
Accepted: 7 September 2021
Published: 9 September 2021

The energy system of the future is a work in progress. Many countries around the world have made a target for their energy system to be completely renewable. How the power system eventually forms, such as the business models, the key players, and its architecture, as well as how it works, will be dependent on the outcomes of trends, forces, regulations, and strategic actions by many diverse players in the energy sector.

Digitalisation of the traditional generation plan and transmission has been proposed. However, there is no existing work in the literature on the digitalisation of the wind turbine generation or solar photovoltaic farms. The future power system will be very complex due to the multiple forces affecting various levels of the system, especially the distribution network level. The systems are composed of thousands of local distribution areas operated by distribution operators (suppliers) on top of the consumers, etc. Thus, although it may not appear too different, the power flows are no longer just one way, going from the bulk power system to the consumer end. In fact, the power flow will be very hard to trace because it can be from one consumer to another, the consumer to the grid, or the grid to the consumer, and some will be mobile/random due to the charging/discharging of electric vehicles.

These types of renewable energy resources (solar PV and wind turbine generation) are incredibly dependent on nature (wind speed, wind direction, temperature, solar irradiation, humidity, etc.). Thus, the outputs are highly stochastic in nature. Data science techniques for handling real-time big data will ideally fit this stream. Furthermore, integrated systems modelling methods and concepts are needed for the study of the self-organisation, complexity, emergent properties, and dynamical behaviour of complex systems for their holistic understanding, management, and development based primarily on neural networks, fuzzy and soft systems/fuzzy cognitive maps, network modelling, and mathematics. Other advanced applications in computational early detection of mastitis and computer-based decision support systems for complex systems are also needed. Due to the scale of the network and the amount of data that need to be digitised, new techniques in data mining and AI approaches are needed in order to analyse and predict the behaviour of these complex systems.

I would like to thank the staff and reviewers for their efforts and input. The task of editing and selecting papers for this collection was found to be both stimulating and rewarding [1–6].

Funding: This research received no external funding.

Institutional Review Board Statement: Not applicable.

Informed Consent Statement: Not applicable.

Conflicts of Interest: The author declares no conflict of interest.

References

1. Bastos, A.F.; Santoso, S. Optimization Techniques for Mining Power Quality Data and Processing Unbalanced Datasets in Machine Learning Applications. *Energies* **2021**, *14*, 463. [CrossRef]
2. Wu, L.; Vanayagamoorthy, G.K.; Gao, J. Online Steady-State Security Awareness Using Cellular Computation Networks and Fuzzy Techniques. *Energies* **2021**, *14*, 148. [CrossRef]
3. Olivieri, C.; de Paulis, F.; Orlandi, A.; Pisani, C.; Giannuzzi, G.; Salvati, R.; Zaottini, R. Estimation of Modal Parameters for Inter-Area Oscillations Analysis by a Machine Learning Approach with Offline Training. *Energies* **2020**, *13*, 6410. [CrossRef]
4. Kim, J.-G.; Lee, B. Automatic P2P Energy Trading Model Based on Reinforcement Learning Using Long Short-Term Delayed Reward. *Energies* **2020**, *13*, 5359. [CrossRef]
5. Hemeida, M.G.; Alkhalaf, S.; Mohamed, A.-A.A.; Ibrahim, A.A.; Senjyu, T. Distributed Generators Optimization Based on Multi-Objective Functions Using Manta Rays Foraging Optimization Algorithm (MRFO). *Energies* **2020**, *13*, 3847. [CrossRef]
6. Al Karim, M.; Currie, J.; Lie, T.-T. Distributed Machine Learning on Dynamic Power System Data Features to Improve Resiliency for the Purpose of Self-Healing. *Energies* **2020**, *13*, 3494. [CrossRef]

MDPI

Article

Optimization Techniques for Mining Power Quality Data and Processing Unbalanced Datasets in Machine Learning Applications

Alvaro Furlani Bastos * and Surya Santoso

Department of Electrical and Computer Engineering, The University of Texas at Austin, Austin, TX 78712, USA; ssantoso@mail.utexas.edu
* Correspondence: alvaro.fbastos@utexas.edu

Abstract: In recent years, machine learning applications have received increasing interest from power system researchers. The successful performance of these applications is dependent on the availability of extensive and diverse datasets for the training and validation of machine learning frameworks. However, power systems operate at quasi-steady-state conditions for most of the time, and the measurements corresponding to these states provide limited novel knowledge for the development of machine learning applications. In this paper, a data mining approach based on optimization techniques is proposed for filtering root-mean-square (RMS) voltage profiles and identifying unusual measurements within triggerless power quality datasets. Then, datasets with equal representation between event and non-event observations are created so that machine learning algorithms can extract useful insights from the rare but important event observations. The proposed framework is demonstrated and validated with both synthetic signals and field data measurements.

Keywords: change detection; data analytics; data mining; filtering; machine learning; optimization; power quality; signal processing; total variation smoothing

Citation: Furlani Bastos, A.; Santoso, S. Optimization Techniques for Mining Power Quality Data and Processing Unbalanced Datasets in Machine Learning Applications. *Energies* **2021**, *14*, 463. https://doi.org/10.3390/en14020463

Received: 24 December 2020
Accepted: 14 January 2021
Published: 16 January 2021

1. Introduction

The application of machine learning algorithms has expanded noticeably in many fields in the last few decades, especially due to the increased power and reduced expense of computation, the growth of field data collection and the advent of novel techniques to process and analyze large datasets. This trend has also been observed in power systems, where most machine learning applications are related to distributed energy resources (such as solar, wind, and storage) and smart grid control. Such examples include the following: load and demand forecasts [1,2], electricity production forecasts [2], solar radiation forecasts [1], wind speed/power forecasts [3], automated control of smart grids [2], management of electric vehicle fleets [1], predictive maintenance [4], fault detection and location [3,5,6] and power quality disturbance classification [3].

Data for machine learning applications in power systems can be acquired from multiple sources. A common source of field data measurements is power quality monitors (PQM), which record instantaneous voltage and current waveforms with a high time resolution (hundreds of samples per cycle). The latest version of these devices allows the addition of precise and synchronized time stamps to the measured data, expanding the suitability of the recorded data to more advanced applications [7]. Traditionally, PQMs employ a limited set of triggering features to detect disturbances within the dataset and, once they have been detected, store a few waveform cycles as individual events [8]. More recently, however, a triggerless data acquisition approach has emerged, where all waveform samples are stored for further analysis.

The main advantage of this approach is that even inconspicuous disturbances are successfully captured [9]; on the other hand, triggerless PQMs generate voluminous datasets and require large data storage capabilities [10,11]. Further, most of the data correspond to the steady-state operation of the power system, whereas only a small part of the recorded data shows disturbances. In other words, the dataset is highly unbalanced, with the steady-state observations heavily outnumbering the disturbance observations, which, as will be discussed later, can cause performance deterioration in most machine learning algorithms. Thus, one of the focuses of this paper is dataset rebalancing, such that the disturbance and non-disturbance classes are equally represented in the input dataset prior to its use by a machine learning algorithm.

1.1. Disturbance Detection in Power System Datasets

Multiple techniques have been proposed over the years for detecting disturbances in PQM data, and they are broadly classified into two categories: in the first category, the trigger mechanism is based on the magnitude of a time series (e.g., overvoltage, overcurrent, signal rate of rise and root-mean-square (RMS) voltage variations) [12] or employs time–frequency and time–scale transformations to decompose the signal into several subbands (e.g., short-time Fourier transform and wavelet transform) [13,14]; the second category is composed of methods based on prominent signal residuals, which are obtained through time-varying mathematical models (e.g., autoregressive (AR) models and Kalman filters) or direct data comparison (e.g., point-by-point or cycle-by-cycle comparison) [15].

It has been shown that these techniques are effective for detecting conspicuous disturbances (i.e., cases where the underlying system event causes transients in the voltage and/or current waveforms) [16]. On the other hand, they are unable to detect most inconspicuous disturbances [17,18], hindering their suitability for the processing of triggerless PQM datasets. Further, they might be sensitive to harmonics, sampling frequency and other user-selected parameters (such as a mother wavelet for the detector based on a wavelet transform). These drawbacks often result in disturbances being missed by the detector, especially those that are very short and/or subtle.

Although waveforms collected by PQMs are valuable assets for power system analysis, these raw measurements might not directly provide useful information for disturbance identification and classification [12]. In fact, various power system events might not cause conspicuous disturbances in the PQM waveforms; instead, they are characterized by an abrupt step change in the RMS voltage profile. Common examples of power system events that belong to this category include capacitor switching de-energizing [19], transformer tap-changing, voltage regulator operation and switching of large loads [13].

Thus, RMS voltage step changes have been proposed as an alternative triggering feature to detect events (both conspicuous and inconspicuous) within PQM datasets [8,12]. This task, however, is complicated by the fact that the magnitude of these RMS voltage step changes is often quite small (even less than 0.5% of the nominal voltage). Moreover, the presence of rapidly varying fluctuations in an RMS voltage profile hinders the detection of RMS voltage step changes. Therefore, prior to being used in the disturbance identification process, the RMS voltage profile must be processed to remove those rapid voltage fluctuations. The desired output of this process is an RMS voltage profile with a high signal-to-noise ratio and sharp edges during the step changes [8], which is another focus of this paper.

1.2. Contributions

This paper proposes a framework for the detection of RMS voltage step changes and rebalancing of highly unbalanced PQM datasets. Its main contributions are as follows: (a) the proposal of a strategy for filtering RMS voltage profiles such that rapidly varying noise is removed or significantly attenuated, whilst preserving the steep edges of the RMS voltage profile caused by switching events; (b) the automatic detection of RMS voltage step changes in the filtered RMS voltage profile, so that both conspicuous and inconspicuous events within a PQM dataset are identified; and (c) the proposal of a framework for rebalancing highly unbalanced PQM datasets.

All optimization problems presented in this paper are implemented and solved in Pyomo [20].

1.3. Article Organization

The remainder of this paper is organized as follows. Section 2 discusses the effects of highly unbalanced datasets in machine learning training and proposes a strategy for rebalancing highly unbalanced PQM datasets. Section 3 presents a literature review on the filtering of RMS voltage profiles and detection of RMS voltage step changes. Section 4 describes the proposed approaches for RMS voltage profile filtering and dataset rebalancing, as well as presenting the PQM datasets analyzed throughout this paper. Section 5 demonstrates the performance of the proposed framework through field data measurements and Section 6 addresses some final considerations.

2. The Problem of Unbalanced Datasets in Machine Learning Training

The main goal of any machine learning algorithm is to learn patterns directly from the data through some computational methods, without relying on a predetermined physical model or some other strong assumptions about the data features. In general, the performance of these algorithms improves as the amount and variability of available samples increase [21,22]. The growth in popularity of machine learning applications is a direct consequence of the rise in big data, as most rule-based models are inadequate to extract insight from such large, complex and ever-changing datasets. Some real-world machine learning applications already in use include such diverse fields as the following [23]:

- Computational finance, for credit scoring and algorithmic trading;
- Image processing and computer vision, for face recognition, motion detection and object detection;
- Computational biology, for tumor detection, drug discovery and DNA sequencing;
- Energy production, for price and load forecasting;
- Automotive, aerospace and manufacturing, for predictive maintenance;
- Natural language processing.

Machine learning methods are broadly classified into two categories: supervised learning, where the algorithm tries to establish a mapping between input features and output targets so that it can be used to predict the output target for future input features; and unsupervised learning, where there is no output target and the goal is to group and interpret the data based only on its input features. As will become clear in the following discussion, the focus of this paper is on supervised learning—either in terms of classification (i.e., the output variable is categorical/discrete) or regression (i.e., the output variable is continuous).

A machine learning application is often divided into three stages—training, validation and testing—with the input dataset split into three corresponding subsets as well [21,22]. Figure 1 represents the general workflow of a typical machine learning application. First, the training set (which usually is the largest of the three subsets) is used to train a machine learning model. The performance of the resulting model is then evaluated using the validation set. If its performance is satisfactory with respect to some metric, the current model is considered as the final version of the machine learning model; otherwise, an iterative loop of successive training and validation stages is executed to incrementally improve the model's predictive power until the desired performance is achieved. This training/validation loop consists of hyper-parameter tuning (if the selected algorithm has any hyper-parameters) or even the selection of an entirely different algorithm. Due to the large number of machine learning algorithms, this step involves some trial-and-error, as there is no one-size-fits-all approach in machine learning (i.e., there is no algorithm that outperforms all other counterpart algorithms for all types of application, datasets size and types of data or desired insights).

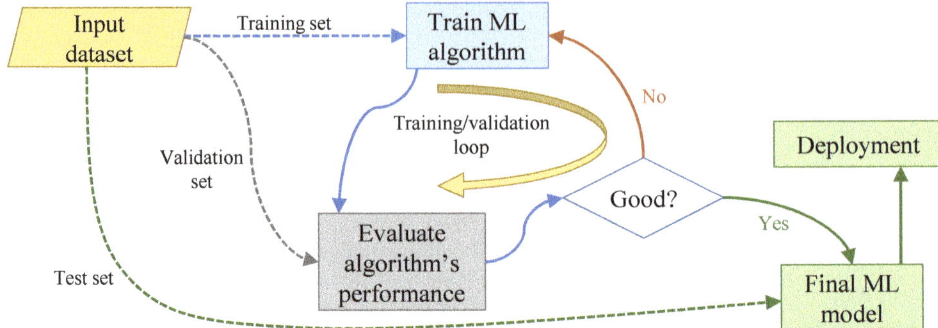

Figure 1. General workflow of a typical machine learning application.

Once the final machine learning model has been selected, the test set is used to produce its performance metrics. It is important to emphasize that the test set should not overlap with the training and validation sets, as the goal in this evaluation step is to estimate the predictive power of the final model on samples that have not been used to fine-tune the model's parameters.

In this paper, we focus on a pre-processing step to be employed prior to any of those three stages in an attempt to improve the performance of the machine learning algorithm. More specifically, our focus is on the handling and processing of highly unbalanced input datasets; i.e., cases in which the observations in the training dataset belonging to one class heavily outnumber the observations in the other class. A general overview of this pre-processing step is shown in Figure 2; the components of the dataset rebalancing block are detailed in Figure 5.

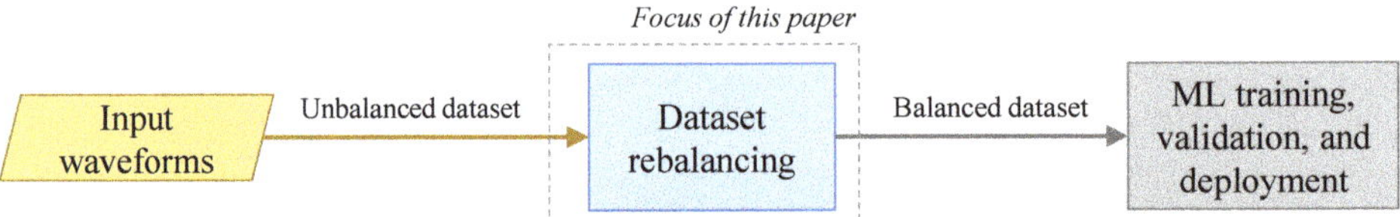

Figure 2. General overview of the work presented in this paper.

Highly unbalanced datasets in machine learning training might influence the model performance and often result in a phenomena called the accuracy paradox. This occurs when the accuracy measure simply reflects the underlying class distribution, rather than learning the actual patterns present in the dataset. Most standard machine learning algorithms are developed under the assumption that the class distributions are roughly balanced. When presented with unbalanced datasets, these algorithms fail to capture the effects of severe class distribution skewness [24], as well as experience difficulties learning the concepts related to the minority class [25].

For example, consider a binary classification problem where the training dataset is composed of 95% of observations for Class 1 and only 5% of observations for Class 2. Most of the machine learning algorithms tend to be biased toward the majority class. If an algorithm classifies a new observation based only on the majority class in the training set (Class 1 in this case), its accuracy would be 95%, which is an excellent value for most practical applications. This approach, however, does not take into account the features of each observation; i.e., there is no actual learning during the training stage, and the final machine learning model is likely to have low predictive accuracy on new observations.

The drawbacks caused by unbalanced datasets might be even worse than is apparent [26]. For example, consider the study presented in [27], where the goal is to predict voltages throughout a distribution network. Not surprisingly, most of the target values in the training dataset are around 1.0 pu, with only a few observations for which the target value is less than 0.95 pu or greater than 1.05 pu. However, prediction accuracy is more important for these scenarios with extreme voltage values (the minority class) than scenarios with voltages around 1.0 pu (the majority class). This difference in prediction accuracy importance is due to the fact that the scenarios with very low or very high voltages are those in which a voltage control device has to operate.

There are multiple practical examples in which class unbalance is quite common and even expected to occur. The minority class often represents rare but important events [25]. A well-known example is represented by the datasets of credit card transactions, where nearly all transactions were authorized by the card holder (not-fraud class), while only a few transactions belong to the fraud class. A similar situation is observed in power system measurements: most of the measurements correspond to steady-state conditions (non-event), while only a few of them are events.

Multiple strategies have been proposed for handling unbalanced datasets, including the following [24,28,29]:

- Collect more data;
- Explore alternative performance metrics, such as the confusion matrix, precision, recall, F-score, Cohen's kappa and receiver operating characteristic (ROC) curves [30];
- Resample the dataset (either through under-sampling or over-sampling, depending on the dataset's initial size);
- Generate synthetic observations;
- Investigate penalized models, where additional costs are imposed on the misclassification of the minority class during training and a higher cost of prediction is associated with rarity [31];
- Reconstruct the training dataset, where the minority observations are identified through anomaly or change detection.

This paper employs the resampling and change detection approaches to construct balanced training datasets. Given an RMS voltage profile, a training dataset is constructed as follows:

1. Partition the input profile into multiple equal-length segments and determine which contain significant changes in the RMS voltage levels; a significant change is defined as an RMS voltage step change greater than a pre-specified threshold, which will be introduced in later sections. Each one of these selected segments corresponds to one observation of the minority class (event) in the training dataset—let n_E denote the number of such observations;
2. Among the segments without a significant change in the RMS voltage level (non-event), randomly select n_E segments to form the majority class (non-event) in the training dataset.

Note that the minority and majority classes are used in the steps above only for consistency with the previous discussion. In fact, the newly created training dataset is evenly balanced between the two classes.

3. The State-of-the-Art

As mentioned in Section 1, this paper focuses on the detection of substantial changes in RMS voltage profiles, so that datasets with a more balanced ratio between events and non-events can be obtained for use in the training and validation stages of a machine learning application pipeline. The most straightforward method to detect such changes in an RMS voltage profile is based on RMS voltage gradients. There are also other alternative detectors proposed in the literature, and these are discussed below.

3.1. The RMS Voltage Gradient Profile Detection Approach

Let the vector $V \in \mathbb{R}^n$ represent an RMS voltage profile with a one-sample time resolution; then, the corresponding RMS voltage gradient profile is defined as

$$\Delta V_k = V_k - V_{k-pN}, \text{ for } k = pN + 1, \ldots, n \tag{1}$$

where N is the number of waveform samples per cycle. The quantity pN controls which RMS voltage values are compared to each other; it is recommended to adopt $p \geq 2$ [13] so that there is at least a one-cycle gap between the waveform samples used to compute V_k and V_{k-pN}. Otherwise, the magnitude of the RMS voltage gradient might be smaller than the true magnitude of the step change when those sets of waveform samples contain a mix of both event and non-event data, possibly causing the event to be undetected [12]. On the other hand, adopting $p \geq 2$ guarantees that any waveform transients lasting less than one cycle will have dissipated and that at least one value in the ΔV profile captures the true magnitude of the step change. The computation of the RMS voltage gradient profile is illustrated in Figure 3 for $p = 2$.

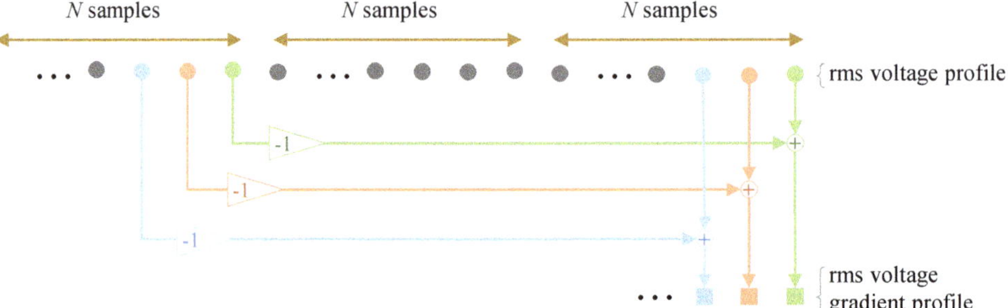

Figure 3. Illustration of the root-mean-square (RMS) voltage gradient profile computation for $p = 2$.

In the RMS voltage gradient approach, an event is detected whenever the following condition is satisfied:

$$|\Delta V_k| > \delta_{\text{step}}, \text{ for } k = pN + 1, \ldots, n \tag{2}$$

where δ_{step} is a pre-specified threshold for the triggering criteria. The chosen value for this threshold has great impacts on the detector's performance, as unsuitable values might cause multiple false positives (δ_{step} is too small) or false negatives (δ_{step} is too large).

In this paper, δ_{step} is selected based on well-known characteristics of power systems; more specifically, we consider switching events that cause the most subtle change in RMS voltage profiles, as described below:

- Voltage regulators are devices that adjust the voltage level by changing the tap positions in an autotransformer. In general, they provide a -10% to $+10\%$ regulation range with 32 steps, where each step represents $\pm 0.625\%$ of the nominal voltage [32].
- Switched capacitor banks cause voltage variations, the magnitudes of which depend on the capacitor bank size and the short-circuit capacity at the bank location. For practical scenarios, the voltage variation falls between 0.36% and 4% of the nominal voltage [19,32–34].

Based on this discussion, we adopted $\delta_{\text{step}} = 0.0018$ pu, which follows the rule of thumb of setting the threshold as half of the minimum-expected step change [35]. This detection technique has been shown to achieve high accuracy, especially in cases where the signal-to-noise ratio of the RMS voltage profile is high (i.e., low noise levels) [12,36]. On the other hand, this detector fails if the RMS voltage profile has high noise levels or it has not been properly filtered.

3.2. Alternative Standard Detector

In the 2015 update, the International Electrotechnical Commission (IEC) added the concept of a rapid voltage change (RVC) to one of its standards [14]. An RVC is defined as an abrupt transition between two RMS voltage values, and its detection is performed as follows:

1. Compute the arithmetic mean of the immediately preceding RMS voltage values:

$$\overline{V}_k = \frac{1}{2f} \sum_{p=k-2f+1}^{k} V_k \tag{3}$$

   where f is the system frequency (either 50 or 60 Hz).
2. Flag a new RMS voltage value as part of an RVC if it deviates from $\overline{V}_k$ by more than a given threshold δ_{RVC}:

$$\left|V_k - \overline{V}_k\right| > \delta_{\text{RVC}} \Longrightarrow \text{Flag as RVC} \tag{4}$$

The RVC threshold δ_{RVC} is set by the user according to the desired application; the standard recommends considering values in the range of 0.01 pu to 0.06 pu. Due to the computation of arithmetic means, this detection approach behaves similarly to linear filtering, which, as discussed in the next section, has the drawback of blurring out the steep edges of the signal.

3.3. RMS Voltage Profile Filtering

The event detectors described previously can exhibit great performance degradation if the RMS voltage profiles are contaminated with noise. In the context of this paper, the following are factors that contribute to noise corruption:

- Noise introduced by the measurement device;
- Varying system frequency, which results in incorrect RMS voltage computations, as N waveform samples do not correspond to an integer number of cycles [36,37];
- Small load variations, which create intermittent variations in the RMS voltage profile and have the potential to hinder the detection of the events of interest.

Thus, a low-pass filtering technique must be applied to the RMS voltage profiles as a pre-processing step [35]. Linear filters, such as a moving average filter, have been shown to be effective in removing or attenuating rapidly varying noise while preserving the slowly varying signal. However, they blur out any steep edges of the signal [8,38,39], such as RMS voltage step changes, making this type of filter unfit for applications based on the detection of switching events [8].

On the other hand, median filters are well-known as suitable options for signals that contain sharp edges [39,40]. The performance of median filters can be further improved through an iterated and multiscale filtering approach, where multiple median filters are applied sequentially from a fine scale (narrow window) to a coarse scale (wide window). The goal of this process is to increase the signal-to-noise ratio at each stage such that the advantages of median filtering can be leveraged at increasingly low noise levels [12,39]. Previous work has compared the performance of single-stage and three-stage median filters applied to RMS voltage profiles around capacitor switching instants. It has been shown that both filters successfully attenuate the signal noise while preserving the RMS step changes in most cases; however, the three-stage median filter provided a faster transition between the steady-state levels prior and posterior to the switching instant [12]. This study also presented scenarios in which median filtering (both single- and three-stage) fails; for example, if the signal varies linearly (i.e., not a constant value) immediately before the step change, median filtering is not able to accurately track the signal.

4. Methodology

As mentioned in Section 1, techniques for properly filtering RMS voltage profiles are one of the main contributions of this paper. This section describes the proposed filtering approach, which is demonstrated through test signals.

4.1. Problem Setup

This subsection presents the data analyzed in the paper (both field measurements and synthetic signals), as well as definitions and formulations that are used in later sections.

4.1.1. Data

Field Measurements

The field measurements analyzed in this study consisted of 28-minute continuous power quality data (voltage and current waveforms) collected at the feeder head of a 25 kV, 60 Hz radial distribution system with multiple parallel feeders. The power quality monitor was installed immediately downstream of the substation transformer, and its sampling frequency was 7.68 kHz (i.e., 128 waveform samples per cycle). The entire monitoring period contained eight major switching events: four capacitor energizing operations and four capacitor de-energizing operations. Further, some relatively large load switching events were also observed, although they had a smaller impact on the RMS voltage profile compared to capacitor switching events.

Synthetic Signals

Synthetic signals were also used in this study because they contained information about the true RMS voltage value without noise contamination at each time instant. The following signals are analyzed in later sections:

- Signal 1: The voltage level in a distribution system was in a quasi-stationary condition at 0.996 pu for 1 second. At that time instant, a capacitor bank was energized, instantaneously increasing the RMS voltage to 1.0 pu. After another 1 second had elapsed, the capacitor bank was de-energized and the RMS voltage level returned to 0.996 pu.
- Signal 2: The voltage level in a distribution system was in a quasi-stationary condition at 1.0 pu for 1 second. At that time instant, the load size connected to the system increased gradually over 1 second, causing the RMS voltage to drop linearly to 0.996 pu. This voltage drop triggered the energizing of a capacitor bank, instantaneously increasing the voltage level back to 1.0 pu. Note: this is the scenario in which median filtering was unable to track the original signal, as mentioned in Section 3.3.

These synthetic signals represented RMS voltage profiles with a half-cycle time resolution, so that each second contained 120 RMS voltage values (for a 60 Hz system). Further, each signal also contained additive noise originating from a normal distribution with zero-mean and standard deviation equal to 0.00025 pu. Figure 4 depicts both synthetic signals.

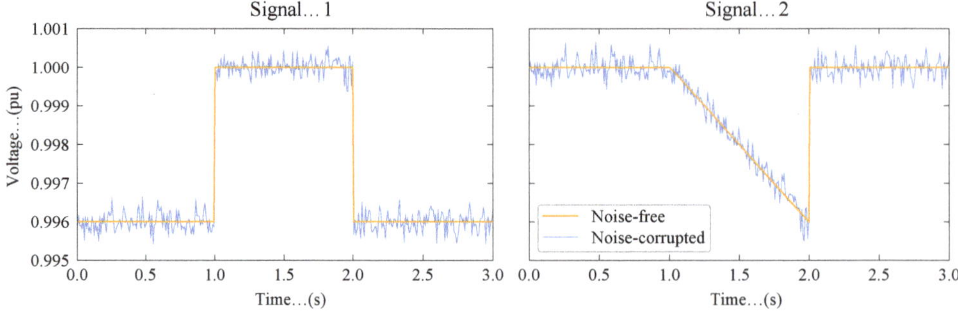

Figure 4. Synthetic signals analyzed throughout this study (both before and after noise addition).

4.1.2. RMS Profile Computation

Let a sampled waveform signal be represented by a vector z; then, its RMS value at instant k, Z_k, is defined as

$$Z_k = \left(\frac{1}{N} \sum_{s=k-N+1}^{k} z_s^2 \right)^{1/2} \tag{5}$$

where N is the number of samples per cycle in the waveform signal. Industrial standards recommend updating an RMS voltage profile every half-cycle ($N/2$ samples) [14,41]; profiles with this time resolution will be indicated as $Z^{(1/2)}$ in the rest of this paper.

On the other hand, computing a new RMS value once every waveform sample becomes available might result in hundreds or thousands of updates per second. The high computational burden involved in this approach is often pointed out as a drawback of having RMS profiles with such a high time resolution [13,14]. However, it has been shown that a recursive approach eliminates such issues [36]. In the recursive approach, the RMS value at instant k is computed as

$$Z_k = \left[Z_{k-1}^2 + \frac{1}{N} \left(z_k^2 - z_{k-N}^2 \right) \right]^{1/2} \tag{6}$$

This recursive approach will be employed throughout the paper wherever RMS profile computation with a high time resolution is necessary.

4.1.3. Vector Norms

For a given vector $z \in \mathbb{R}^n$, its l_1-norm (Manhattan norm) and l_2-norm (Euclidean norm) are defined according to Equations (7) and (8), respectively:

$$\|z\|_1 = \sum_{i=1}^{n} |z_i| \tag{7}$$

$$\|z\|_2 = \left(\sum_{i=1}^{n} z_i^2 \right)^{1/2} \tag{8}$$

Note that in the following sections, the squared Euclidean norm $\|z\|_2^2$ is preferred over $\|z\|_2$ in order to avoid the square root operator.

4.2. Proposed Approach

Figure 5 depicts an overview of the PQM dataset rebalancing framework proposed in this paper. First, the input voltage waveforms were converted into the corresponding RMS voltage profiles (Section 4.1.2), which were filtered to remove/attenuate additive noise (Section 4.3). The filtered RMS voltage profiles were segmented into fixed-length, non-overlapping windows (in this study, we set each window length to 1 s). Each one of these segments was classified as an event or non-event, using the RMS voltage gradient profile approach that was introduced in Section 3.1. After all segments were classified into one of the two categories, dataset rebalancing was performed as described in Section 2. Finally, the resulting dataset could be used for the training/validation of machine learning algorithms.

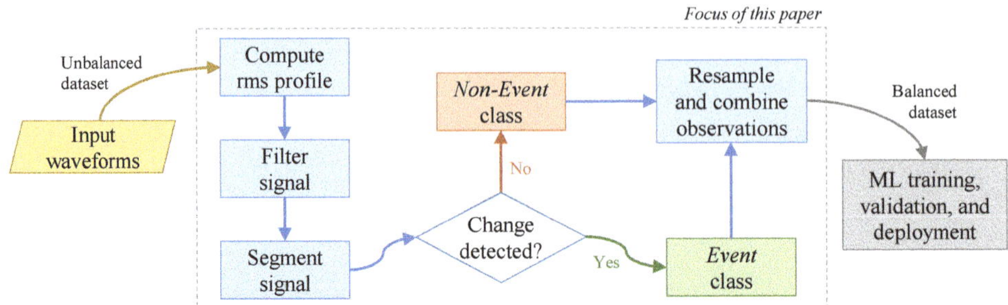

Figure 5. Overview of the power quality monitor (PQM) dataset rebalancing framework proposed in this paper.

4.3. Data Filtering

This section describes the filtering of time series through optimization techniques. Consider a signal represented by the vector $x \in \mathbb{R}^n$, where each coefficient x_i represents the signal value sampled at the i-th time instant and the sampling interval is fixed. Without loss of generality, it is often assumed that the signal does not vary too rapidly for most of the time, as was the case for the signals analyzed in this study, so that $x_i \approx x_{i+1}$.

As commonly observed in field measurements, the signal x is corrupted by an additive noise ν, i.e., $x_{\text{cor}} = x + \nu$. Note that x_{cor} is observable by measurement devices, whereas the true underlying signal x is unknown. The additive noise ν can be modeled based on known characteristics of the process under study; however, for generality, it will be assumed that it follows an unknown distribution, has a small amplitude and varies much more rapidly than the signal x [42].

The objective of time series filtering is to produce an estimate x^* of the original signal x, given the corrupted signal x_{cor}; this process is also called signal reconstruction or denoising. The reconstructed signal x^* should be similar to the corrupted signal and smooth; i.e., the rapidly varying noise is removed or significantly attenuated. The closeness between the corrupted and reconstructed signals is often measured with respect to the l_2-norm, and a penalty function ϕ is used to assess the non-smoothness of the reconstructed signal. Thus, this signal filtering problem can be formulated as a convex vector optimization problem [42], as follows:

$$x^* = \underset{\hat{x} \in \mathbb{R}^n}{\operatorname{argmin}}\, \mathbf{F}(\hat{x}, x_{\text{cor}}) \tag{9}$$

where the objective function

$$\mathbf{F}(\hat{x}, x_{\text{cor}}) = \begin{bmatrix} \|\hat{x} - x_{\text{cor}}\|_2 \\ \phi(\hat{x}) \end{bmatrix} \tag{10}$$

is a vector. Its first component, $F_1 = \|\hat{x} - x_{\text{cor}}\|_2$ represents a measure of fit or consistency between the corrupted and estimated signals, whereas the second component, $F_2 = \phi(\hat{x})$, measures the roughness or lack of smoothness of the estimate $\hat{x}$. The function $\phi : \mathbb{R}^n \to \mathbb{R}$ is convex and often given as some norm. Note, however, that F_1 and F_2 do not need to be measured with respect to the same norm, and this fact will be exploited later to produce better estimates for x^*. In problems involving l_2-norms, it is common practice to consider the corresponding squared norms [42], so that the nonlinearities caused by square roots are removed from the problem formulation; thus, we will adopt $F_1 = \|\hat{x} - x_{\text{cor}}\|_2^2$.

The formulation presented in Equation (9) corresponds to a multi-objective optimization problem, where each component can be interpreted as different scalar objectives. The goal is to minimize each one of these components; however, they represent competing objectives, and a decrease in F_1 is accompanied by an increase in F_2 and vice-versa.

A standard approach for solving such optimization problems is called scalarization or regularization, where the objective function in Equation (9) is reformulated as

$\boldsymbol{\lambda}^{\mathrm{T}}\mathbf{F}(\hat{x}, x_{\mathrm{cor}}) = \lambda_1\|\hat{x} - x_{\mathrm{cor}}\|_2^2 + \lambda_2\phi(\hat{x})$ for any weight vector $\boldsymbol{\lambda} > 0$ [42]. Note that $\boldsymbol{\lambda}^{\mathrm{T}}\mathbf{F}(\hat{x}, x_{\mathrm{cor}})$ is scalar-valued and convex, since it is a weighted sum of convex functions [43]. Therefore, the reformulated problem is an ordinary scalar convex optimization problem, which can be solved easily.

The weight vector $\boldsymbol{\lambda}$ has a great influence on the filtering process as it controls the smoothness level of the output signal, and choosing a suitable value is critical to achieving the desired level of noise removal [44]. In general, each choice of $\boldsymbol{\lambda}$ results in a different estimate x^* [42]. Let $\boldsymbol{\lambda} = [1, \delta]^{\mathrm{T}}$, for some $\delta > 0$; as δ varies over $[0, \infty)$, the solution of the equivalent scalar optimization problem traces out the optimal trade-off curve (or Pareto curve) between minimizing each component F_1 and F_2 separately. Figure 6 depicts a typical Pareto curve for a bi-criterion vector optimization problem, where values of components F_1 and F_2 are plotted on the horizontal and vertical axes, respectively.

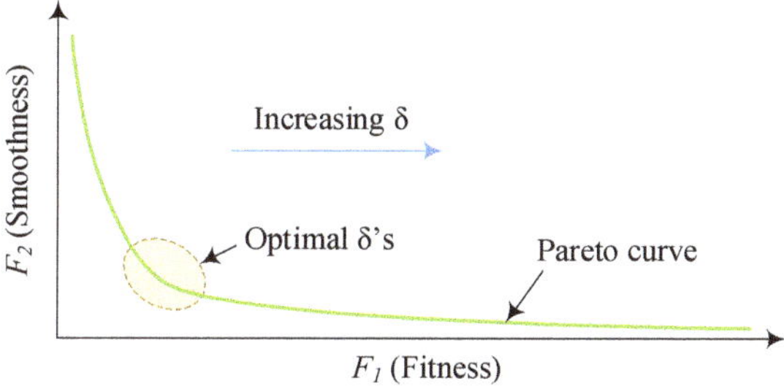

Figure 6. Typical Pareto curve for a bi-criterion vector optimization problem.

For any δ, the slope of the Pareto curve represents the local optimal trade-off between the two objectives: if the slope is steep, small changes in F_1 are accompanied by large changes in F_2, and vice-versa [42]. In other words, a Pareto curve allows us to determine how large one of the objectives must be in order to have the other one be small. Thus, the filtering of signals with a low signal-to-noise ratio (high noise levels) requires a larger δ [44].

In the extremes of a Pareto curve, we have the following interpretation:

- $\delta = 0$: there is no penalty associated with the roughness of the output signal; thus, no smoothing is performed and $x^* = x_{\mathrm{cor}}$. This scenario corresponds to the endpoint at the left in the Pareto curve, and it represents the smallest possible value of F_1 without any consideration of F_2.
- $\delta \to \infty$: a stronger emphasis is placed on the smoothness of the output signal, at the expense of disregarding the similarity between the corrupted and estimated signals; for a sufficiently large δ, x^* becomes a constant signal. This scenario corresponds to the endpoint at the right in the Pareto curve, and it represents the smallest possible value of F_2 without any consideration of F_1.

Choosing a suitable δ is a compromise between F_1 and F_2. In practice, its value is chosen empirically by analyzing the Pareto curve and selecting a value such that a small decrease in one objective is accompanied by a small increase in the other objective [42]. The δ values that satisfy this requirement form the knee of the Pareto curve.

In the next sections, we present different strategies for quantifying the smoothness of the filtered signal; i.e., we present formulations for the component $F_2 = \phi(\hat{x})$ of the objective function.

4.3.1. Quadratic Smoothing

The most straightforward roughness measure of a signal is given in terms of the sum of squares of differences. The quadratic smoothing function is defined as

$$F_2 = \phi_{\mathrm{quad}}(\hat{x}) = \sum_{i=1}^{n-1}(\hat{x}_{i+1} - \hat{x}_i)^2 = \|D\hat{x}\|_2^2 \tag{11}$$

where $D \in \mathbb{R}^{(n-1)\times n}$ is the bidiagonal matrix

$$D = \begin{bmatrix} -1 & 1 & 0 & \cdots & 0 & 0 & 0 \\ 0 & -1 & 1 & \cdots & 0 & 0 & 0 \\ \vdots & \vdots & \vdots & \ddots & \vdots & \vdots & \vdots \\ 0 & 0 & 0 & \cdots & -1 & 1 & 0 \\ 0 & 0 & 0 & \cdots & 0 & -1 & 1 \end{bmatrix}_{(n-1)\times n} \tag{12}$$

and represents an approximation to the first-order differentiation operator. As $\phi_{\text{quad}}(\hat{x})$ is defined in terms of a l_2-norm, its squared value is used in the optimization problem, as discussed previously.

The estimate x^* is the solution to the following unconstrained scalar optimization problem:

$$\underset{\hat{x}\in\mathbb{R}^n}{\text{minimize}} \|\hat{x} - x_{\text{cor}}\|_2^2 + \delta\|D\hat{x}\|_2^2 \tag{13}$$

where $\delta > 0$ parametrizes the optimal trade-off curve between $\|\hat{x} - x_{\text{cor}}\|_2^2$ and $\|D\hat{x}\|_2^2$. This formulation corresponds to a quadratic problem, which can be solved very efficiently [42].

Figure 7a shows the Pareto curve for $\delta \in [0, 500]$ for the synthetic signal 1 defined in Section 4.1.1, where it can be observed that $\delta \approx 2$ is the optimal weight. Figure 7b depicts three smoothed signals on the optimal trade-off curve:

- $\delta = 0.2$ (under-filtering): the weight associated with the output signal roughness is too small; although the steep edges in the signal are preserved, there is almost no reduction in the signal noise.
- $\delta = 2$ (optimal): this scenario represents the optimal trade-off between corrupted and estimated signals similarity and noise reduction; however, the noise level in the filtered signal is still quite high.
- $\delta = 100$ (over-filtering): an excessive weight is placed on the signal smoothness, resulting in over-filtering; the similarity between the corrupted and estimated signals is rather low.

Figure 8 shows the filtering results for the synthetic signal 2, and the discussion presented above is also valid for this test case.

This analysis shows that quadratic smoothing either removes the rapidly varying noise or preserves the steep signal edges, but not both; in fact, quadratic smoothing behaves as a low-pass filter. Thus, this technique is not suitable for the category of signals analyzed in this paper.

4.3.2. Total Variation Smoothing

Given the limitations of quadratic smoothing discussed previously, this section describes a smoothing function that is effective at removing/attenuating the signal noise, while still preserving the steep edges of the original signal [45,46]. In this case, the signal smoothness is measured according to the following function:

$$F_2 = \phi_{\text{tv}}(\hat{x}) = \sum_{i=1}^{n-1} |\hat{x}_{i+1} - \hat{x}_i| = \|D\hat{x}\|_1 \tag{14}$$

which is called the total variation of $\hat{x} \in \mathbb{R}^n$. Note that, compared to ϕ_{quad} in Equation (11), ϕ_{tv} is not squared, as it is given in terms of a l_1-norm and there are no square root terms to be removed.

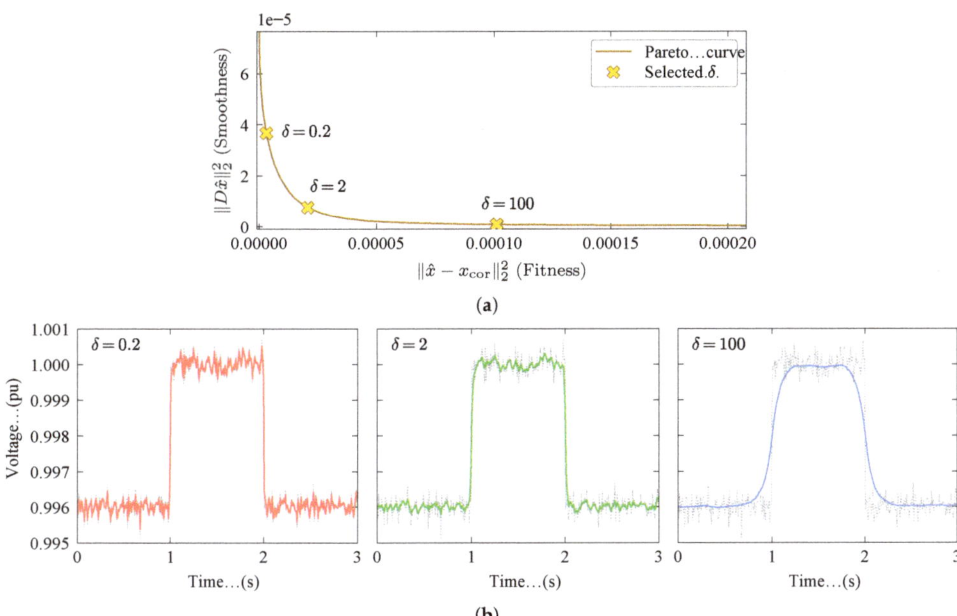

Figure 7. Results of filtering the corrupted synthetic signal 1 through quadratic smoothing. (**a**) Pareto curve. (**b**) Estimated signals representing under-filtering ($\delta = 0.2$), optimal filtering ($\delta = 2$) and over-filtering ($\delta = 100$).

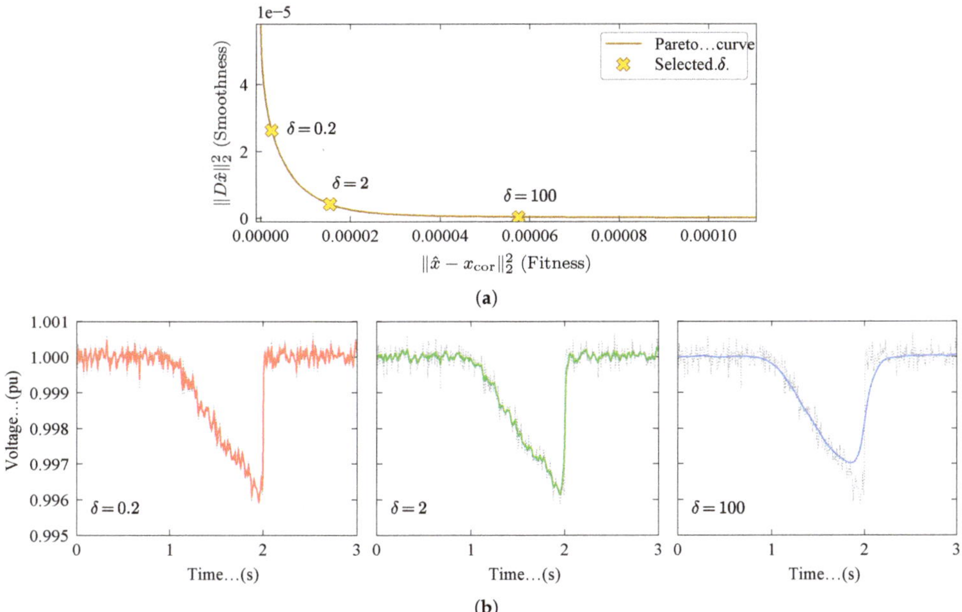

Figure 8. Results of filtering the corrupted synthetic signal 2 through quadratic smoothing. (**a**) Pareto curve. (**b**) Estimated signals representing under-filtering ($\delta = 0.2$), optimal filtering ($\delta = 2$) and over-filtering ($\delta = 100$).

The estimate x^* is the solution of the following unconstrained scalar optimization problem:

$$\underset{\hat{x}\in\mathbb{R}^n}{\text{minimize}} \|\hat{x} - x_{\text{cor}}\|_2^2 + \delta\|D\hat{x}\|_1 \tag{15}$$

The optimization problem in Equation (15) cannot be easily solved because the l_1-norm is non-differentiable [47]. The following problem reformulation is based on [43]. First, for simplicity, we introduce a new variable $y_i = \hat{x}_{i+1} - \hat{x}_i, \forall i = 1, \ldots, n-1$, so that $\phi_{\text{tv}}(\hat{x}) = \sum_{i=1}^{n-1} |y_i|$.

Let $y_i = y_i^+ - y_i^-, \forall i = 1, \ldots, n-1$, where y_i^+ and y_i^- are variables constrained to be nonnegative. It can be shown that these two variables cannot be simultaneously nonzero; i.e., at least one of the variables y_i^+ and y_i^- is zero for each index i. Therefore,

$$y_i = \begin{cases} y_i^+, & \text{if } y_i \geq 0 \quad (y_i^+ \geq 0, y_i^- = 0) \\ -y_i^-, & \text{if } y_i < 0 \quad (y_i^+ = 0, y_i^- > 0) \end{cases} \implies |y_i| = y_i^+ + y_i^- \tag{16}$$

By replacing $|y_i|$ in Equation (15), the following alternative formulation is obtained:

$$\begin{aligned} \underset{\hat{x},y,y^+,y^-\in\mathbb{R}^n}{\text{minimize}} \; & \|\hat{x} - x_{\text{cor}}\|_2^2 + \delta \sum_{i=1}^{n-1} (y_i^+ + y_i^-) \\ \text{subject to} \; & y_i = \hat{x}_{i+1} - \hat{x}_i, \quad i = 1, \ldots, n-1 \\ & y_i = y_i^+ - y_i^-, \quad i = 1, \ldots, n-1 \\ & y_i^+ \geq 0, \quad i = 1, \ldots, n-1 \\ & y_i^- \geq 0, \quad i = 1, \ldots, n-1 \end{aligned} \tag{17}$$

which is a constrained, convex and differentiable optimization problem.

Figure 9 demonstrates the filtering of synthetic signal 1 through total variation smoothing. Figure 9a shows the Pareto curve for $\delta \in [0, 5]$, where it can be observed that $\delta \approx 0.004$ is the optimal weight. Figure 9b depicts three smoothed signals on the optimal trade-off curve:

- $\delta = 0.0002$ (under-filtering): the weight associated with the output signal roughness is too small, meaning that there is almost no reduction in the signal noise.
- $\delta = 0.004$ (optimal): this scenario represents the optimal trade-off between corrupted and estimated signal similarity and noise reduction.
- $\delta = 0.2$ (over-filtering): an excessive weight is placed on the signal smoothness, resulting in over-filtering; due to the large penalty associated with variations in the signal, the magnitude of the step change in the filtered signal is much smaller than the magnitude of the true step change.

Figure 10 shows the filtering results for the synthetic signal 2, and the discussion presented above is also valid for this test case. Further, unlike median filtering, total variation smoothing was able to track this piecewise linear signal.

This analysis shows that total variation smoothing exhibits great performance in noise reduction without blurring the sharp transitions of the original signal, as long as the weight δ has been properly selected.

4.3.3. Quadratic vs. Total Variation Smoothing

As discussed in the previous sections, total variation smoothing shows better performance in the filtering of RMS voltage profiles when compared to quadratic smoothing. In this section, we explore and compare the characteristics of these two smoothing operators in order to justify the superiority achieved by total variation smoothing.

Both ϕ_{quad} and ϕ_{tv}, which were defined in Equations (11) and (14), respectively, assign large penalty costs to rapidly varying $\hat{x}$. However, the quadratic smoothness function assigns a relatively small penalty to small values of $|\hat{x}_{i+1} - \hat{x}_i|$ [48]. For example, if $|\hat{x}_{i+1} - \hat{x}_i| = 0.001$, then the penalties assigned by the quadratic and total variation smoothness functions are 10^{-6} and 10^{-3}, respectively. In other words, the quadratic smoothing operator tolerates some variation in the filtered signal, whereas the total variation smoothing operator is subject to a much larger penalty if such signal variations exist, meaning that it enforces $|\hat{x}_{i+1} - \hat{x}_i| \approx 0$ for almost all i's.

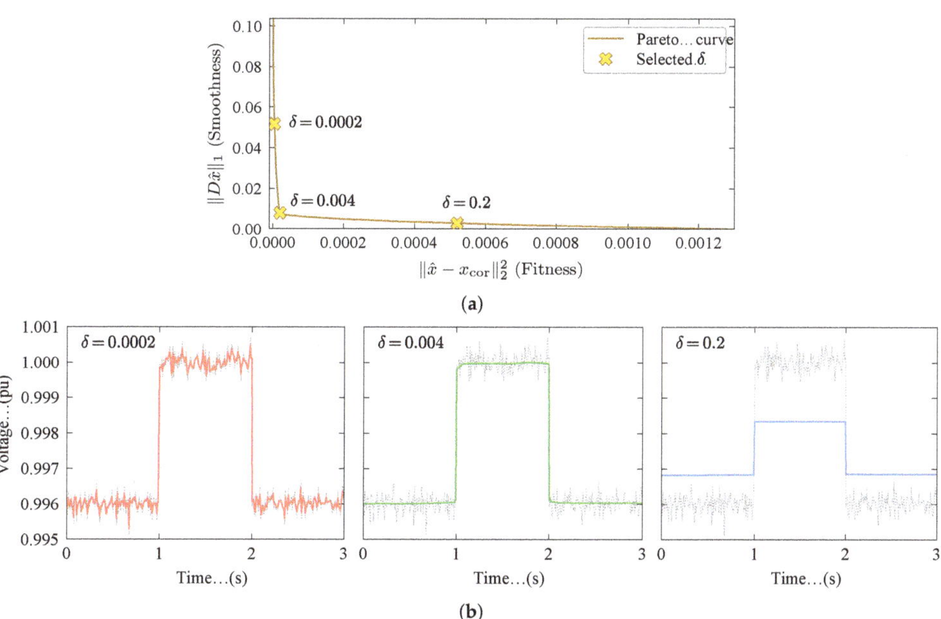

Figure 9. Results of filtering the corrupted synthetic signal 1 through total variation smoothing. (**a**) Pareto curve. (**b**) Estimated signals representing under-filtering ($\delta = 0.0002$), optimal filtering ($\delta = 0.004$) and over-filtering ($\delta = 0.2$).

In general, the following characteristics are observed in the solutions of optimization problems with penalty functions [42]:

- l_2-norm penalty: $||D\hat{x}||_2$ has many non-zero small entries and relatively few larger ones;
- l_1-norm penalty: $||D\hat{x}||_1$ has many zero or very small entries and more larger ones.

The optimization problem scalarized with an l_1-norm is a heuristic for finding a solution in which $||D\hat{x}||_1$ is sparse. As D represents an approximation to the first-order differentiation operator, total variation smoothing is biased toward solutions in which the filtered signal is linear or piecewise linear.

This behavior can be observed in Figure 11, which depicts the histogram of $|x_{i+1} - x_i|$ for the filtered signals computed in Section 4.3.1 (Figure 7b, quadratic smoothing with $\delta = 2$) and Section 4.3.2 (Figure 9b, total variation smoothing with $\delta = 0.004$), respectively. As expected, quadratic smoothing allows some $|\hat{x}_{i+1} - \hat{x}_i|$ to be greater than zero, which correspond to the smooth transition around the steep edges of the original signal. On the other hand, almost all $|\hat{x}_{i+1} - \hat{x}_i|$ in Figure 11b are approximately zero, except for two values that correspond to the two step changes present in the original signal.

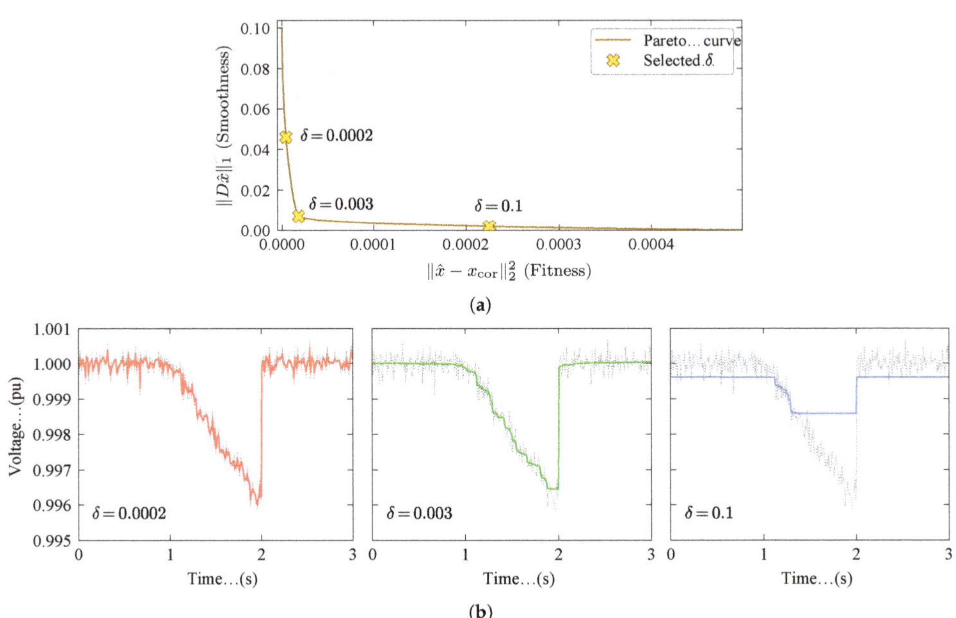

Figure 10. Results of filtering the corrupted synthetic signal 2 through total variation smoothing. (**a**) Pareto curve. (**b**) Estimated signals representing under-filtering ($\delta = 0.0002$), optimal filtering ($\delta = 0.003$) and over-filtering ($\delta = 0.1$).

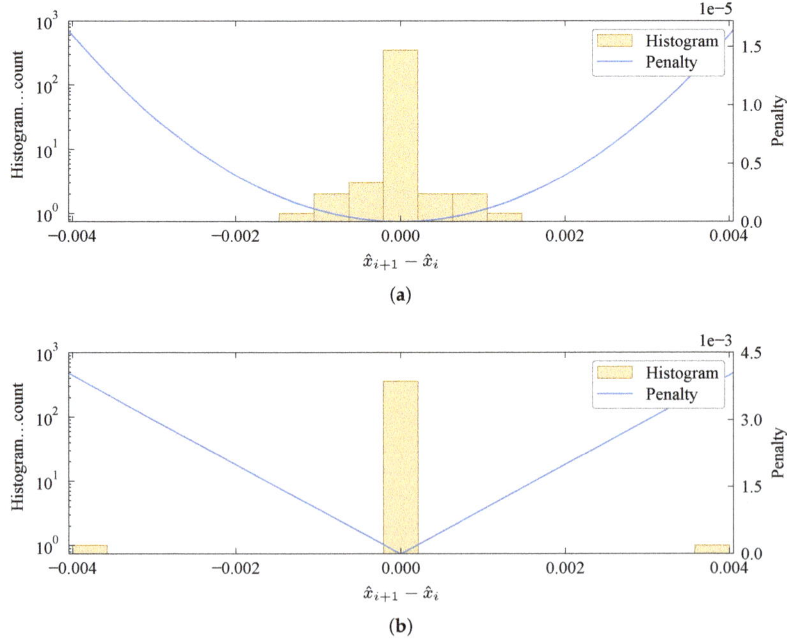

Figure 11. Histogram of the first derivative amplitudes for the filtered synthetic signal 1 using the optimal δ value for each scenario. (**a**) Quadratic smoothing with $\delta = 2$. (**b**) Total variation smoothing with $\delta = 0.004$.

5. Results

This section demonstrates the application of the proposed framework (i.e., total variation smoothing) using field data collected at the feeder head of a 25 kV radial distribution system, which is described in Section 4.1.1. Based on the results in Section 4.3.2, we adopted $\delta = 0.0035$. Figure 12a shows the unfiltered and filtered RMS voltage profiles for the entire 28-minute measurement interval, whereas Figure 12b shows the corresponding RMS voltage gradient profiles. Using the triggering threshold $\delta_{\text{step}} = 0.0018$ pu (Section 3.1), all RMS voltage step changes were successfully detected without any false positives. The root causes of the detected RMS voltage step changes were capacitor de-energizing (events 1, 2, 5 and 6) and capacitor energizing (events 3, 4, 7 and 8). Further, the unfiltered RMS voltage gradient profile did not create any false positives either; however, the gradient values were much larger compared to the filtered cases (as large as 0.0015 pu), indicating that the unfiltered profile might create false positives for some datasets.

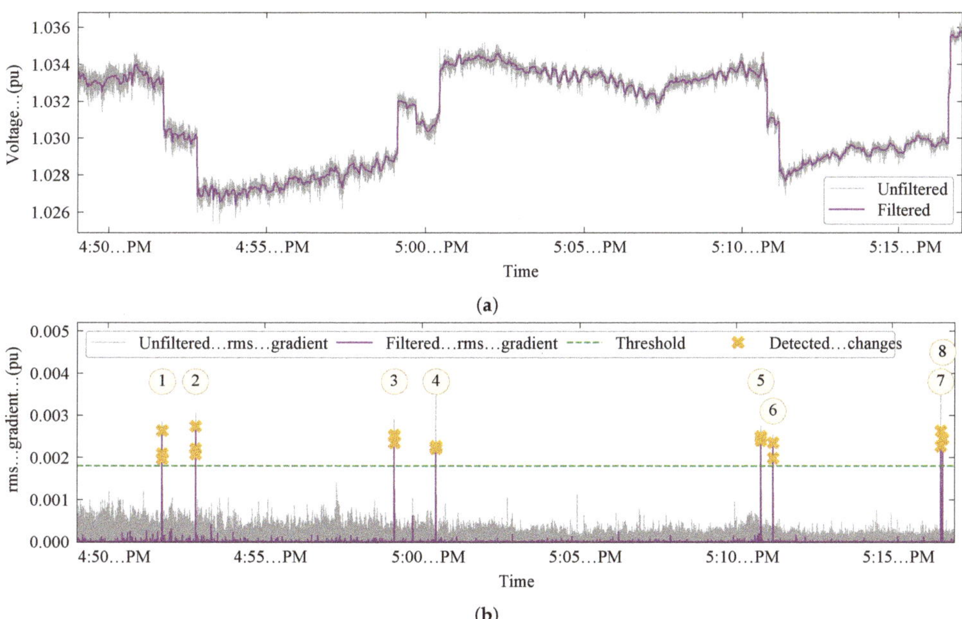

Figure 12. Results from the field data. (**a**) Unfiltered and filtered RMS voltage profiles; the filtered profile was obtained through total variation smoothing with $\delta = 0.0035$. (**b**) Unfiltered and filtered RMS voltage gradient profiles, where the numbers in circles represent event IDs.

Detailed views of the unfiltered and filtered RMS voltage profiles are shown in Figure 13 for four scenarios: capacitor de-energizing, capacitor energizing, load energizing and steady-state. Note that the filtered profile did not contain rapidly varying noise and its step changes were not affected, as initially desired. The unfiltered RMS voltage profile in Figure 13b contained a spike immediately after the RMS voltage step change, which was due to high-frequency transients in the voltage waveform caused by a capacitor energizing operation. On the other hand, total variation smoothing successfully removed this spike. This is an important advantage of using the filtered profile, as the magnitude of the step change might be one of the features employed by the machine learning algorithm (the magnitude given by the unfiltered profile is about 50% larger than the correct value).

Both unfiltered and filtered RMS voltage profiles were segmented into non-overlapping 1 s windows and classified as an event or non-event, as described in Section 4.2. Table 1 shows the distribution of classes before and after the rebalancing of the PQM dataset.

Table 1. Class distribution for the PQM dataset before and after rebalancing.

	Before Rebalancing	After Rebalancing
Majority class (*Event*)	1672 (99.52%)	8 (50%)
Minority class (*Non-Event*)	8 (0.48%)	8 (50%)

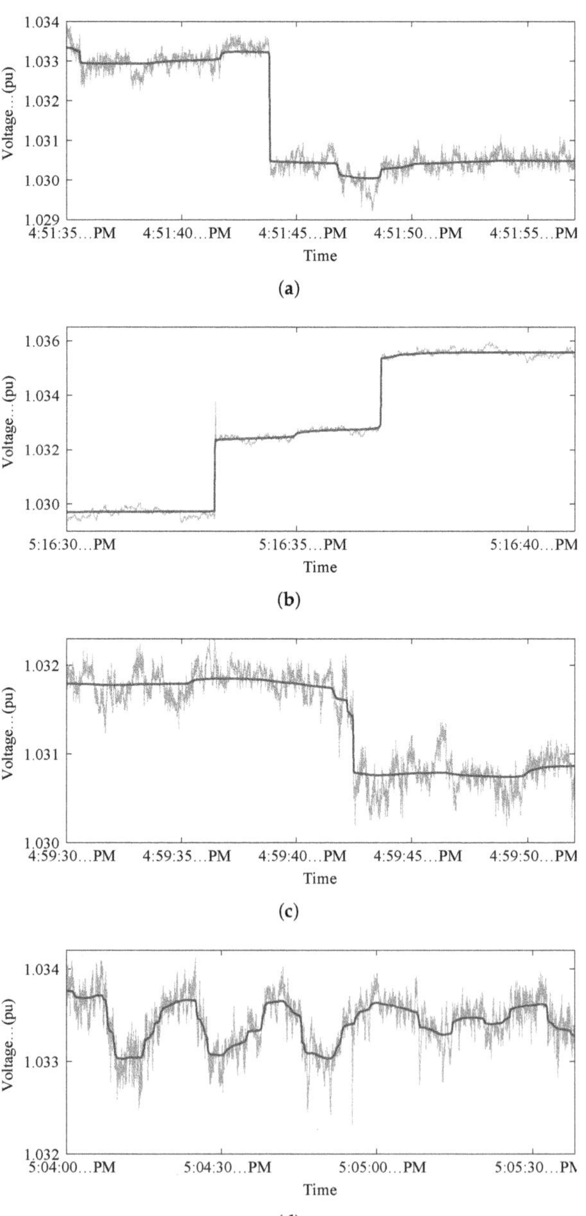

Figure 13. Detailed view of the unfiltered and filtered RMS voltage profiles for the field data. (**a**) Capacitor de-energizing (event 1). (**b**) Successive capacitor energizing (events 7 and 8). (**c**) Load energizing (between events 3 and 4). (**d**) Steady-state (between events 4 and 5).

Before dataset rebalancing, less than 0.5% of the observations in the input dataset corresponded to power system disturbances; in this case, machine learning algorithms are very unlikely to be able to extract any useful information about the minority class. On the hand, the dataset was perfectly balanced using the framework proposed in this paper. It should be noted, however, that the rebalanced dataset contained only 16 observations, which is often considered too small for successfully training machine learning algorithms. One solution would be to select more observations for the majority class, as long as the resulting dataset does not become highly unbalanced. Another solution consists of collecting more PQM data; the field data considered in this paper represent only 28 minutes of measurement, whereas utilities have access to much longer measurement intervals (days, weeks or even months).

6. Conclusions

The RMS voltage profile filtering proposed in this paper was shown to be robust for removing/attenuating rapidly varying signal noise without blurring out the RMS voltage step changes due to switching events. By combining filtering and step change detection techniques, both conspicuous and inconspicuous events present in a PQM dataset can be successfully identified. Detecting such events is the basis for rebalancing highly unbalanced PQM datasets, consequently improving the performance of machine learning algorithms that use these datasets in their training and validations stages.

As observed in Figures 7b, 8b, 9b and 10b, the parameter δ has a great effect on the RMS voltage profile filtering process. Further, the optimal value for δ depends on the noise level present in the signal; i.e., scenarios with a higher signal-to-noise ratio (low noise level) require a lower δ value. Therefore, the optimal δ value adopted in this paper might not be the most suitable choice for field measurements collected at other locations, as the signal-to-noise ratio might be different.

Future research directions include the development of techniques for automatically determining the optimal δ for each dataset. For example, for a given RMS voltage profile, such techniques would first analyze only a short segment of the profile for multiple δ values in order to construct the Pareto curve. Then, the optimal δ would be the value corresponding to the knee of the Pareto curve, as shown in Figure 6. Once this optimal value has been determined, the whole RMS voltage profile would be filtered through total variation smoothing. It is important to emphasize that optimization applications based on the Pareto curve in all fields—and not only power systems—empirically determine the optimal δ by visually inspecting the Pareto curve. Thus, a technique for automatically determining this value would represent a meaningful contribution.

Author Contributions: The conceptualization, methodology, software development, data analysis and writing of the manuscript was done by A.F.B.; S.S. was the advisor mentoring the project. All authors have read and agreed to the published version of the manuscript.

Funding: This research received no external funding.

Conflicts of Interest: The authors declare no conflict of interest.

Abbreviations

The following abbreviations are used in this manuscript:

AR	Autoregressive
ML	Machine learning
PQM	Power quality monitor
pu	Per unit
RMS	Root-mean-square
ROC	Receiver operating characteristic
RVC	Rapid voltage change

References

1. Mosavi, A.; Salimi, M.; Ardabili, S.F.; Rabczuk, T.; Shamshirband, S.; Varkonyi-Koczy, A.R. State of the art of machine learning models in energy systems: A systematic review. *Energies* **2019**, *12*, 1301. [CrossRef]
2. Kumar, N.M.; Chand, A.A.; Malvoni, M.; Prasad, K.A.; Mamun, K.A.; Islam, F.R.; Chopra, S.S. Distributed energy resources and the application of AI, IoT, and blockchain in smart grids. *Energies* **2020**, *13*, 5739. [CrossRef]
3. Perez-Ortiz, M.; Jimenez-Fernandez, S.; Gutierrez, P.A.; Alexandre, E.; Hervas-Martinez, C.; Salcedo-Sanz, S. A review of classification problems and algorithms in renewable energy applications. *Energies* **2016**, *9*, 607. [CrossRef]
4. Bastos, A.F.; Santoso, S. Condition monitoring of circuit-switchers for shunt capacitor banks through power quality data. *IEEE Trans. Power Deliv.* **2019**, *34*, 1499–1507. [CrossRef]
5. Ananthan, S.N.; Bastos, A.F.; Santoso, S.; Chirapongsananurak, P. Model-based approach integrated with fault circuit indicators for fault location in distribution systems. In Proceedings of the IEEE Power and Energy Society General Meeting, Atlanta, GA, USA, 4–8 August 2019; pp. 1–5.
6. Ananthan, S.N.; Bastos, A.F.; Santoso, S. Novel system model-based fault location approach using dynamic search technique. *IET Gener. Transm. Distrib.* **2021**. [CrossRef]
7. Bastos, A.F.; Santoso, S.; Freitas, W.; Xu, W. SynchroWaveform measurement units and applications. In Proceedings of the IEEE Power and Energy Society General Meeting, Atlanta, GA, USA, 4–8 August 2019; pp. 1–5.
8. Bastos, A.F.; Lao, K.W.; Todeschini, G.; Santoso, S. Novel moving average filter for detecting rms voltage step changes in triggerless PQ data. *IEEE Trans. Power Deliv.* **2018**, *33*, 2920–2929. [CrossRef]
9. Xu, W. Experiences on Using Gapless Waveform Data and Synchronized Harmonic Phasors. In *Panel Session in IEEE Power and Energy Society General Meeting*; Technical Report; IEEE Power & Energy Society: Piscataway, NJ, USA, 2015.
10. Silva, L.; Kapisch, E.; Martins, C.; Filho, L.; Cerqueira, A.; Duque, C.; Ribeiro, P. Gapless power-quality disturbance recorder. *IEEE Trans. Power Deliv.* **2017**, *32*, 862–871. [CrossRef]
11. Li, B.; Jing, Y.; Xu, W. A generic waveform abnormality detection method for utility equipment condition monitoring. *IEEE Trans. Power Deliv.* **2017**, *32*, 162–171. [CrossRef]
12. Bastos, A.F.; Freitas, W.; Todeschini, G.; Santoso, S. Detection of inconspicuous power quality disturbances through step changes in rms voltage profile. *IET Gener. Transm. Distrib.* **2019**, *13*, 2226–2235. [CrossRef]
13. Bollen, M.; Gu, I. *Signal Processing of Power Quality Disturbances*; Wiley: Hoboken, NJ, USA, 2006.
14. IEC. *IEC Electromagnetic Compatibility: Testing and Measurements Techniques—Power Quality Measurement Methods*; Standard 61000-4-30; IEC: London, UK, 2015.
15. Bastos, A.F.; Santoso, S. Universal waveshape-based disturbance detection in power quality data using similarity metrics. *IEEE Trans. Power Deliv.* **2020**, *35*, 1779–1787. [CrossRef]
16. Santoso, S.; Powers, E.J.; Grady, W.M.; Parsons, A.C. Power quality disturbance waveform recognition using wavelet-based neural classifier—Part 1: Theoretical foundation. *IEEE Trans. Power Deliv.* **2000**, *15*, 222–228. [CrossRef]
17. Bastos, A.F.; Lao, K.W.; Todeschini, G.; Santoso, S. Accurate identification of point-on-wave inception and recovery instants of voltage sags and swells. *IEEE Trans. Power Deliv.* **2019**, *34*, 551–560. [CrossRef]
18. Bastos, A.F.; Santoso, S.; Todeschini, G. Comparison of methods for determining inception and recovery points of voltage variation events. In Proceedings of the IEEE Power and Energy Society General Meeting, Portland, OR, USA, 5–10 August 2018; pp. 1–5.
19. Bastos, A.F.; Santoso, S. Identifying switched capacitor relative locations and energizing operations. In Proceedings of the IEEE Power and Energy Society General Meeting, Boston, MA, USA, 17–21 July 2016; pp. 1–5.
20. Hart, W.E.; Laird, C.D.; Watson, J.P.; Woodruff, D.L.; Hackebeil, G.A.; Nicholson, B.L.; Siirola, J.D. *Pyomo—Optimization Modeling in Python*, 2nd ed.; Springer Science & Business Media: Berlin/Heidelberg, Germany, 2017; Volume 67.
21. Bishop, C.M. *Pattern Recognition and Machine Learning*, 1st ed.; Springer: Berlin/Heidelberg, Germany, 2006.
22. Duda, R.O.; Hart, P.E.; Stork, D.G. *Pattern Classification*, 2nd ed.; Wiley-Interscience: Hoboken, NJ, USA, 2000.
23. Gron, A. *Hands-On Machine Learning with Scikit-Learn and TensorFlow: Concepts, Tools, and Techniques to Build Intelligent Systems*, 1st ed.; O'Reilly Media, Inc.: Newton, MA, USA, 2017.
24. He, H.; Garcia, E.A. Learning from imbalanced data. *IEEE Trans. Knowl. Data Eng.* **2009**, *21*, 1263–1284.
25. Batista, G.E.A.P.A.; Prati, R.C.; Monard, M.C. A study of the behavior of several methods for balancing machine learning training data. *ACM SIGKDD Explor. Newsl.* **2004**, *6*, 20–29. [CrossRef]
26. Torgo, L.; Ribeiro, R.P. *Precision and Recall for Regression*; Discovery Science; Springer: Berlin/Heidelberg, Germany, 2009; pp. 332–346.
27. Bastos, A.F.; Santoso, S.; Krishnan, V.; Zhang, Y. Machine learning-based prediction of distribution network voltage and sensor allocation. In Proceedings of the IEEE Power and Energy Society General Meeting, Montreal, QC, Canada, 2–6 August 2020; pp. 1–5.
28. Brownlee, J. 8 Tactics to Combat Imbalanced Classes in Your Machine Learning Dataset. 2015. Available online: https://machinelearningmastery.com/tactics-to-combat-imbalanced-classes-in-your-machine-learning-dataset/ (accessed on 20 December 2020).
29. Chawla, N.V. Data Mining for Imbalanced Datasets: An Overview. In *Data Mining and Knowledge Discovery Handbook*; Maimon, O., Rokach, L., Eds.; Springer: Boston, MA, USA, 2005; pp. 853–867.

30. Tamayo, S.C.; Luna, J.M.F.; Huete, J.F. On the use of weighted mean absolute error in recommender systems. In Proceedings of the Workshop on Recommendation Utility Evaluation, Dublin, Ireland, 9 September 2012; pp. 24–26.
31. Torgo, L.; Ribeiro, R.P.; Pfahringer, B.; Branco, P. SMOTE for Regression. In *Lecture Notes in Computer Science, Proceedings of the Progress in Artificial Intelligence, Azores, Portugal, 9–12 September 2013*; Springer: Berlin/Heidelberg, Germany, 2013; pp. 378–389.
32. Short, T.A. *Electric Power Distribution Handbook*; CRC Press: Boca Raton, RL, USA, 2003.
33. IEEE. *IEEE Guide for Application of Shunt Power Capacitors*; Standard 1036; IEEE: Piscataway, NJ, USA, 2010.
34. Bastos, A.F.; Biyikli, L.; Santoso, S. Analysis of power factor over correction in a distribution feeder. In Proceedings of the IEEE Power and Energy Society Transmission and Distribution Conference and Exposition, Dallas, TX, USA, 3–5 May 2016; pp. 1–5.
35. Gustafsson, F. *Adaptive Filtering and Change Detection*; John Wiley & Sons: Hoboken, NJ, USA, 2000.
36. Bastos, A.F.; Santoso, S. Root-mean-square profiles under varying power frequency: Computation and applications. In Proceedings of the IEEE Power and Energy Society General Meeting, Atlanta, GA, USA, 4–8 August 2019; pp. 1–5.
37. Bastos, A.F.; Kim, T.; Santoso, S.; Grady, W.M.; Gravois, P.; Miller, M.; Kadel, N.; Schmall, J.; Huang, S.H.; Blevins, B. Frequency retrieval from PMU data corrupted with pseudo-oscillations during off-nominal operation. In Proceedings of the North American Power Symposium, Tempe, AZ, USA, 11–14 April 2021; pp. 1–6.
38. Smith, S. *The Scientist and Engineer Guide to Digital Signal Processing*; California Tech. Pub.: San Diego, CA, USA, 1997.
39. Castro, E.A.; Donoho, D.L. Does median filtering truly preserve edges better than linear filtering? *Ann. Stat.* **2009**, *37*, 1172–1206. [CrossRef]
40. Huber, P.J. Robust estimation of a location parameter. *Ann. Math. Stat.* **1964**, *35*, 73–101. [CrossRef]
41. IEEE. *IEEE Guide for Voltage Sag Indices*; Standard 1564; IEEE: Piscataway, NJ, USA, 2014.
42. Boyd, S.; Vandenberghe, L. *Convex Optimization*, 7th ed.; Cambridge University Press: Cambridge, UK, 2009.
43. Bertsimas, D.; Tsitsiklis, J.N. *Introduction to Linear Optimization*, 1st ed.; Athena Scientific: Belmont, MA, USA, 1997.
44. Selesnick, I.W.; Bayram, I. Total Variation Filtering. 2010. Available online: https://eeweb.engineering.nyu.edu/iselesni/lecture_notes/TV_filtering.pdf (accessed on 20 December 2020).
45. Rudin, L.I.; Osher, S.; Fatemi, E. Nonlinear total variation based noise removal algorithms. *Phys. D Nonlinear Phenom.* **1992**, *60*, 259–268. [CrossRef]
46. Little, M.A.; Jones, N.S. Sparse Bayesian step-filtering for high-throughput analysis of molecular machine dynamics. In Proceedings of the IEEE International Conference on Acoustics, Speech and Signal Processing, Dallas, TX, USA, 14–19 March 2010; pp. 4162–4165.
47. Chambolle, A. An algorithm for total variation minimization and applications. *J. Math. Imaging Vis.* **2004**, *20*, 89–97.
48. Strong, D.; Chan, T. Edge-preserving and scale-dependent properties of total variation regularization. *Inverse Probl.* **2003**, *19*, S165–S187. [CrossRef]

MDPI

Article

Online Steady-State Security Awareness Using Cellular Computation Networks and Fuzzy Techniques

Lili Wu [1,*], Ganesh K. Venayagamoorthy [2,3,*] and Jinfeng Gao [1]

1 School of Electrical Engineering, Zhengzhou University, Zhengzhou 450001, China; jfgao@zzu.edu.cn
2 Real-Time Power and Intelligent Systems Laboratory, Holcombe Department of Electrical and Computer Engineering, Clemson University, Clemson, SC 29634, USA
3 Eskom Centre of Excellence in HVDC Engineering, University of KwaZulu-Natal, Durban 4041, South Africa
* Correspondence: wuli1988@126.com (L.W.); gkumar@ieee.org (G.K.V.)

Abstract: Power system steady-state security relates to its robustness under a normal state as well as to withstanding foreseeable contingencies without interruption to customer service. In this study, a novel cellular computation network (CCN) and hierarchical cellular rule-based fuzzy system (HCRFS) based online situation awareness method regarding steady-state security was proposed. A CCN-based two-layer mechanism was applied for voltage and active power flow prediction. HCRFS block was applied after the CCN prediction block to generate the security level of the power system. The security status of the power system was visualized online through a geographic two-dimensional visualization mechanism for voltage magnitude and load flow. In order to test the performance of the proposed method, three types of neural networks were embedded in CCN cells successively to analyze the characteristics of the proposed methodology under white noise simulated small disturbance and single contingency. Results show that the proposed CCN and HCRFS combined situation awareness method could predict the system security of the power system with high accuracy under both small disturbance and contingencies.

Keywords: steady-state security assessment; situation awareness; cellular computational networks; load flow prediction; contingency; fuzzy system

Citation: Wu, L.; Venayagamoorthy, G.K.; Gao, J. Online Steady-State Security Awareness Using Cellular Computation Networks and Fuzzy Techniques. *Energies* **2021**, *14*, 148. https://doi.org/10.3390/en14010148

Received: 19 November 2020
Accepted: 22 December 2020
Published: 30 December 2020

1. Introduction

With the development of grid interconnection, the structure of modern power systems is expanding. It is constantly developing in the direction of high voltage, long distance, and large capacity, and becoming more complex. At the same time, the proliferation of highly permeable renewable energy sources, such as wind and solar energy, makes the electricity market impose loads on the grid in a more unpredictable, uncontrollable, and dynamic way. It is difficult to predict power grid information with an increasingly large geographic area and more dynamic load. Because of that, it is insurmountable for controllers to see the full picture of the power grid situation under a fault or contingency. Therefore, fast, accurate, and predictive estimation of the system security status has become a major concern for dispatchers.

Power system security estimation problems can be classified in dynamic security analysis and static security analysis [1]. At present, due to the long sampling time of supervisory control and data acquisition (SCADA) system, online security analysis is mainly conducted from the perspective of static security analysis, regarding voltage, current, active, and reactive power security, etc. However, the extensive deployment of a phasor measurement unit (PMU) provides a possible solution for fast online security situation awareness. Since dynamic security analysis is based on a steady-state initial value, a fast online updated static state can be used as an initial state of dynamic security analysis. Even static state tracking which is fast enough can be considered as a dynamic security analysis [2].

Various research studies have been conducted on steady-state security awareness using traditional methods. Literatures [3,4] have evaluated the voltage and load flow security using the situation awareness method. Xiao et al. [3] proposed a situation awareness method based on a static voltage security region of a large power system integrated with wind farms. Netto et al. [4] presented an efficient voltage security region construction tool using probabilistic reliability evaluation to solve a situational awareness problem. This probability-based voltage security assessment algorithm provides richer visual information about the system safety level and has the potential for real-time application. Sun et al. [5] proposed a steady-state operation situation awareness method based on the dynamic power flow method of the active distribution network to study the operation state of the power grid under time changes. The traditional methods can supply a fair result for the system situation, but most of them require the exact system model and are highly computational.

Wide applications of artificial intelligence (AI)-based power system security analysis reveal that AI technology is an effective tool for power systems model building, time saving during power flow computation, and online situation awareness. Fuzzy logic is an intelligent algorithm with natural rules which is closer to human thought than traditional logic systems. A particle swarm optimization combined K-means fuzzy algorithm was addressed in [6] for the power system security assessment. Both static and transient security could be classified as secure or insecure under given states and outages. The fuzzy logic clustering technique was adopted in [7] by Matos and so forth to evaluate global multi-contingency steady-state security. Literatures [8–10] proposed a fuzzy logic-based contingency ranking method instead of the conventional performance index approach to overcome the mask problems for power system static security analysis. Marannino et al. [10] proposed a neuro-fuzzy method for the voltage collapse risk classification. Halilčević et al. [11] used fuzzy membership functions of power system elements to estimate the system security level online. Later, Halilčević et al. [12] used the deterministic and fuzzy inference method to continuously estimate the security, adequacy, and reliability of power system current operation. Kalyani et al. [13] generated synchronized phasor measurements to construct a neuro-fuzzy network for online voltage security monitoring. Zhao et al. [14] proposed a hierarchical model for survival situation awareness using variable fuzzy set technology to estimate system survivability. Various applications of fuzzy logic-based power system security analysis reveal that fuzzy technology is a highly promising tool for translating the operator's linguistic experience to executable machine language which can make the operator aware of the security state of power networks. The use of the fuzzy set theory of variables improves the accuracy and objectivity of the evaluation results.

Besides fuzzy logic, other AI methods have been applied in literatures [15–19] regarding power system security awareness. Fan et al. [15] proposed a data-driven system voltage prediction model based on the generalized regression neural network to dig in-depth power system operation big data to enhance system situation awareness. Literature [16] proposed a real-time safety assessment tool based on PMU and a decision tree to estimate potential safety hazards after failure: Voltage amplitude fluctuation, temperature limit violation, voltage stability, and transient stability. Literature [17] formulated post-outage reactive power flow analysis as a nonlinear constrained optimization problem of a bounded network to be solved by the genetic algorithm (GA). Literature [18] assessed the static security of the power system using an enhanced radial basis function (RBF) neural network. Literature [19] proposed a novel steady-state contingency screening method combining the feed-forward neural network (FFNN) and the fast Fourier transform (FFT). The effectiveness of the AI methods has been verified through selected research cases and predetermined schemes. However, because of the time-consuming problem during online neural network training, the above AI-based methods are not suitable for online application.

CCN is a distributed scalable neural network architecture composed of a computing element (neural network or other) in each cell, which is suitable for describing complex nonlinear network systems whose actual model is not available, and learning its dynamic characteristics in time and space [20]. As a distributed dynamic recurrent neural network, the CCN architecture has the advantage of high scalability, effective nonlinear modeling, and easy computation parallelized. The CCN has good performance in solving the problems of transient stability prediction [21], load flow inferencing [22], state estimation [23], dynamic state prediction [24], wide area measurement (WAM) [25], and situation awareness (SA) [26] using measurements from PMUs. In the previous research by the author [27,28], different kinds of neural network-based CCN were applied as an effective tool for power system state estimation. In literature [27], the multi-layer perceptron (MLP)-based CCN was applied to bus voltage prediction. MLP is a traditional neural network which is simple and easy to train. Even though it has not taken advantage of contextual information, the results are acceptable for voltage prediction. In literature [28], recurrent neural network (RNN)-based CCN was applied to state estimation. The results showed that the introduction of the CCN technique to the power system state prediction made it possible for online security analysis application, as the distribute structure of CCN could estimate the operation state with less time consumed. This paper proposed an online situation awareness method considering power system static security using echo state network (ESN)-based CCN and fuzzy logic. Different from MLP and other RNN neural networks which are hard to converge during the learning process, ESN is an effective tool for prediction even based on the simplest line regression training method. To validate the validity of the proposed ESN-based CCN method, MLP- and RNN-based CCN were also applied in this publication to compare with the results in literatures [27,28].

This paper is organized as follows. Section 2 introduces the design of the situation awareness system using the proposed method. The situation awareness was realized through 4 levels (perception level, comprehension level, projection level, and visualization level). Section 3 shows the perception and comprehension levels. In the design of the comprehension level, a two-layer CCN-based state prediction is proposed. Section 4 focuses on the projection level with the design of a hierarchical cellular rule-based fuzzy system (HCRFS)-based system security assessment. Section 5 shows system security visualization utilizing web-based computer language. Section 6 provides the discussion and results under small disturbance and contingency. Finally, Section 7 presents conclusions and suggests potentially promising future work in this field.

2. System Architecture

For modern intelligent power system, fast and accurate situational awareness is particularly important for power system security. When contingency occurs, it can provide an effective judgment basis for the operator in the control room the first time, and avoid wrong operation, missed operation, or delayed operation, which may lead to cascade failures, and even system blackout. The research of the power system situational awareness technology is still in its infancy. It mainly improves the power grid perception ability through information integration, overall control strengthening of power grid, system reliability enhancement, and operator misoperation decrease. Power system situation awareness is to accurately and effectively grasp power grid security situation through three levels: Perception, comprehension, and projection. This paper implemented a CCN and HCRFS combined online situation awareness method regarding power system steady-state security. The proposed situation awareness architecture is shown in Figure 1.

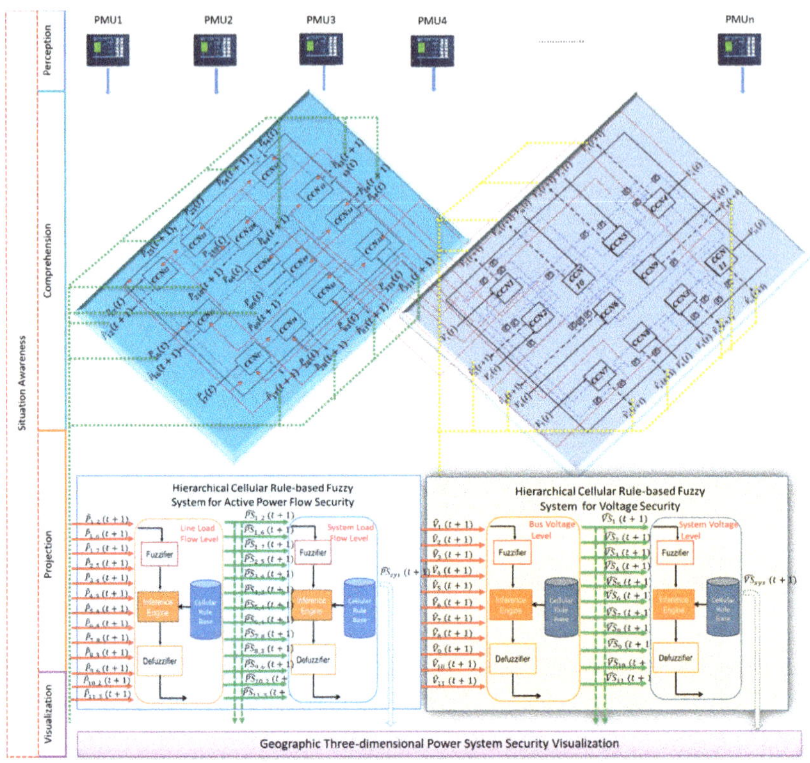

Figure 1. The cellular computation network (CCN)- and hierarchical cellular rule-based fuzzy system (HCRFS)-based power system security situation awareness.

Perception level: 12-bus power system model built in real-time digital simulation system (RTDS) benchmark to generate synchronized power system measurements from PMU.

Comprehension level: CCN-based two-layer online state prediction and estimation using PMU data.

Projection level: A HCRFS-based power system voltage and power security level assessment using prediction states.

Visualization level: A scheme to dynamically visualize the voltage and load flow situation in geographic environment is proposed using the forecasting system security levels.

As shown in Figure 1, in the perception process, PMUs were applied on a 12-bus power system to generate synchronized data. In the comprehension part, the voltage magnitude and load flow were predicted with a two-layer CCN-based mechanism using PMU data. In the projection step, the predicted voltage magnitude and active power were used to assess the power system security level using a HCRFS-based mechanism. Finally, buses and the system security level were displayed geographically two-dimensionally by the data visualization tools. The system architecture shown in Figure 1 is illustrated in detail in Sections 3–5.

3. Perception and Comprehension

The perception and comprehension levels of the proposed situation awareness technology regarding steady-state power system security are illustrated in this section.

3.1. PMU Based Data Generation of 12-Bus Benchmark (Perception)

As an early tentative study, a 12-bus power system was applied to test the steady-state power system security. The 12-bus power system model was built in a real-time digital simulation system (RTDS) benchmark. PMUs were deployed on each bus to generate synchronized power system measurements, like voltage magnitude, voltage angle, current magnitude, and current angle. Voltage violation and load overflow under contingency are common static security problems. In order to fully use the PMU data, the bus voltage and line current measurements were used to predict the bus voltage violation and load overflow.

The selected 12-bus power system had different types of generators and loads, and the network size was suitable for the initial CCN application. The 12-bus platform included 4 generators (Generator G_1 was connected to an infinity bus). The generators' and loads' active power capacity are shown in Table 1, while other details of the test power system can be seen in [29]. As shown in Table 2, there were 13 types of component contingencies in the 12-bus system: 8 transmission line contingencies, 2 transformer outages, and 3 generator trips.

Table 1. The 12-bus power system parameters.

Load/Gen. No.	Active Power/MW
Generator G_1 (infinity Bus)	289
Generator G_2	500
Generator G_3	300
Generator G_4	400
Load on Bus_2	280
Load on Bus_3	320
Load on Bus_4	320
Load on Bus_5	100
Load on Bus_6	440

Table 2. The 12-bus power system line information.

Line No.	From Bus	To Bus	Rating/MW
$Line_{1\_2}$	1	2	250
$Line_{1\_6}$	1	6	250
$Trans_{1\_7}$	1	7	1000
$Line_{2\_5}$	2	5	250
$Line_{3\_4\_1}$	3	4	250
$Line_{3\_4\_2}$	3	4	250
$Line_{4\_5}$	4	5	250
$Line_{4\_6}$	4	6	250
$Line_{7\_8}$	7	8	500
$Trans_{8\_3}$	8	3	1000
G_2	10	2	700
G_3	11	3	500
G_4	9	6	500

3.2. CCN-Based Two-Layer State Prediction (Comprehension)

In the comprehension process of the proposed situation awareness mechanism, a two-layer CCN-based method was proposed for state prediction. In Figure 1, the voltage magnitude of each bus is predicted with the top right CCN layer using PMU data, while the load flow of each transmission line is forecasted based on the top left CCN panel utilizing PMU data and the prediction from the voltage layer. ESN was applied in each cell of the two-layer CCN.

From Figure 1, regarding the voltage prediction layer, there were 11 buses that had been simulated, except the infinity bus of the 12-bus power system. ESN was implemented in each cell representing each of the 11 buses. The relationship of each cell in CCN was a direct mapping of the connections of each bus of the test power system. In each cell, the one-step-ahead voltage magnitude prediction of the bus is defined as below:

$$\left|\widehat{V}_i(t+1)\right| = f\left\{|V_i(t)|,\ \theta_i(t),\ \left|\widehat{V_n}(t)\right|\right\} \tag{1}$$

where $|V_i(t)|$, $\theta_i(t)$ are the voltage, magnitude, and angle of bus i at time t. $\left|\widehat{V_n}(t)\right|$ is the one step delayed voltage prediction values of the neighbors that are connected to bus i.

In the load flow prediction layer, ESN was implemented in each cell representing each line. From Table 2, there are 13 lines in the 12-bus power system. The relationship of each cell in CCN was a direct mapping of the connections of each line of the test power system. In each cell, the predicted active load flow of line j is $\widehat{P}_j(t+1)$ which is shown in Equation (2).

$$\widehat{P}_j(t+1) = f\left\{\left|\widehat{V}_i(t)\right|,\ \theta_j(t), I_j(t)\right\} \tag{2}$$

where $\left|\widehat{V}_i(t)\right|$ is the time-delayed voltage prediction values of bus i generated from Equation (1). $\theta_j(t)$ is the current angle of line j at time t from PMUs. $I_j(t)$ is the line current magnitude at time t of line j.

3.3. The Online Learning of the ESN in Each Cell

ESN is one type of recurrent neural network which uses a dynamic reservoir to simulate the nonlinear relationship between the input and the output.

In an ESN with K inputs, N units in reservoir, and L outputs, the reservoir states are updated following the equation below:

$$X(i+1) = f_{res}\left(W_{in}U(i+1) + WX(i) + W_{fb}Y(i)\right) \tag{3}$$

where W_{in} is the weight between the input layer and reservoir units, W is the weight in the reservoir layer and the output layer, while W_{fb} is the feedback weight. f_{res} is the active function of the reservoir layer (usually the logistic sigmoid or the tanh function). $X(i)$ is the reservoir state, $U(i)$ is the K dimension input signal, and $Y(i)$ is the L dimension output signal. The output is obtained from Equation (4):

$$Y(i+1) = f_{out}(W_{out}(U(i+1), X(i+1), Y(i))) \tag{4}$$

where weights W_{out} is the readout weight matrix between the reservoir layer and f_{out} is the output activation function (typically the identity or a sigmoid).

In order to obtain the W_{out}, W_{in} and W were randomly initialized and Equations (3) and (4) were activated with input and output signals. After that, the W_{out} readout weights matrix could be trained by the line regression method in Equation (5) [30]:

$$W_{out} = (pinv(M) * T)^{Trans} \tag{5}$$

where M is $[(U(i+1),\ X(i+1),\ Y(i)]$ and $pinv(M)$ is the pseudo-inverse of M. T is the inverted of the output active function $fout^{-1}(Y(i+1))$. *Trans* is transpose.

The online training of ESN using Equations (3)–(5) in each cell of the CCN is shown in Figure 2. In Figure 2, there are two modes (training mode and prediction mode) in each cell of the two-layer CCN model. For example, in the voltage prediction layer, the prediction mode worked for continuous voltage prediction using updating weights $W_{out}(k+1)$. The training mode, which was a mirror reflection of the ESN structure in the prediction mode, was activated if the mean square error (MSE) was larger than the expected tolerance. Each ESN/cell was trained with the line regression method in Equation (5) using a dynamic database which was updated with prediction data and target value.

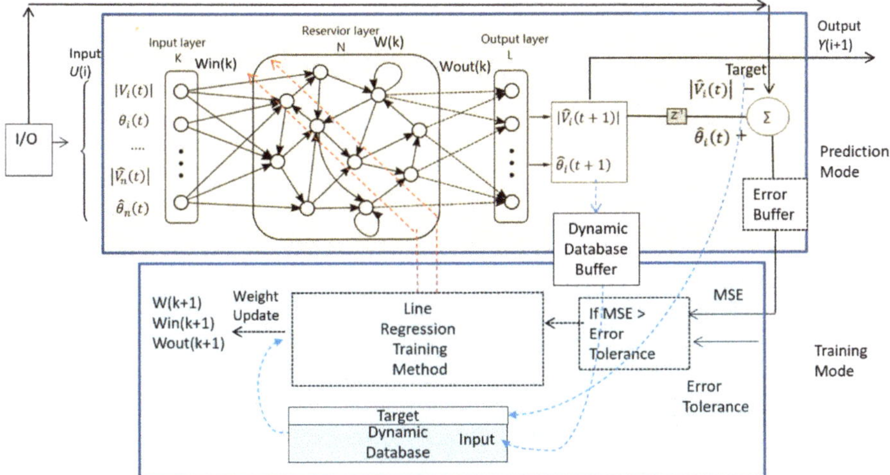

Figure 2. The online training of the echo state network (ESN) in each cell.

In the training mode of Figure 2, the fitness function is the MSE between the prediction value and real value of all the k training data points. The MSE is an effective measurement for forecasting the error of a prediction method in statistics.

$$MSE = \frac{1}{k}\sum_{i=1}^{k}|Act_i - Pred_i|^2 \quad (6)$$

It is defined as the mean of the square of the error between the actual value and prediction value. Where Act_i is the actual value and $Pred_i$ is the prediction value, and k is the number of the data points.

4. Projection

This section shows the projection process which was developed using an HCRFS-based power system security-level classification. The predicted voltage magnitude and active power from the comprehension part were used to assess the power system security level here.

4.1. Voltage Security Assessment

The voltage may violate beyond its limitation under disturbances such as load variations. A definition of voltage collapse, instability, and security introduced by IEEE [31] concluding the power system voltage stability may be threatened in the presence of a variety of single or multiple contingencies. Real-time voltage fluctuation can be performed using the security index [32] below:

$$Index_v(t) = \sum_{i=1}^{p} \omega_i.(\Delta|V_i(t)|)^m \quad (7)$$

where $\Delta|V_i(t)| = |V_i(t)| - |V_{li}|$, $\Delta|V_i(t)|$ is the voltage magnitude violation at time t. $|V_i(t)|$ is voltage magnitude of load bus i at time t. $|V_{li}|$ is the voltage magnitude limit of load bus i. p is the load bus number. ω is weight factor and m is exponent factor.

4.2. Active Power Flow Security Assessment

A possible way to express the real-time security level of overload is the power security index:

$$Index_{MW}(t) = \sum_{j=1}^{k} \gamma_j \cdot (R_j(t))^n \tag{8}$$

where $R_j(t) = (P_j(t))/P_{lj}$, $R_j(t)$ is the active power overload ratio at time t. $P_j(t)$ is the branch j load at time t, and P_{lj} is the overload limit. k is the number of branches. γ is weight factor and n is exponent factor.

The above Equations (7) and (8) denote that the voltage magnitude violation $\Delta|Vi\ (t)|$ and active power rating ratio $R_j(t)$ are potential variables to formulize system security indices. The proposed HCRFS-based security assessment was performed via the predicted voltage magnitude violation $\Delta\left|\widehat{V}_i(t+1)\right|$ and active power rating ratio $\widehat{R}_j(t+1)$.

4.3. HCRFS Based System Security Assessment

From Figure 1, the voltage security assessment was designed via predicted voltage magnitude $\Delta\left|\widehat{V}_i(t+1)\right|$. Meanwhile, load flow security analysis was realized with an active power rating ratio, $\widehat{R}_j(t+1)$. The fuzzy controller transformed the expert knowledge into an automatic control strategy through fuzzification, inference engine, rule base, and defuzzification structure mode. In each designed HCRFS, there were two layers. The first layer was a single bus or line security assessment; meanwhile, the second layer was system level security analysis.

From the HCRFS for voltage security in Figure 1, it is clear that the input was $\Delta\left|\widehat{V}_i(t+1)\right|$, which was the predicted voltage magnitude violation of 11 buses, and the output was the system voltage security level $\widehat{VS}_{sys}(t+1)$. The proposed HCRFS fuzzy index for voltage security was performed by:

$$\widehat{VS}_i(t+1) = fuzzy\left(\Delta\left|\widehat{V}_i(t+1)\right|\right) \tag{9}$$

$$\widehat{VS}_{sys}(t+1) = fuzzy\left(\widehat{VS}_i(t+1)\right) \tag{10}$$

$$\Delta\left|\widehat{V}_i(t+1)\right| = \left|\widehat{V}_i(t+1)\right| - |V_{li}| (i = 1, 2 \ldots .11) \tag{11}$$

where $\widehat{VS}_i(t+1)$ is the voltage security assessment of bus i. $\left|\widehat{V}_i(t+1)\right|$ is the first step ahead predicted voltage magnitude of bus i from CCN which is shown in Equation (1). $|V_{li}|$ is the voltage magnitude limit of load bus i.

The proposed HCRFS-based overload security assessment which is shown in the left bottom panel of Figure 1 can be formulized below:

$$\widehat{PS}_j(t+1) = fuzzy\left(\widehat{R}_j(t+1)\right) \tag{12}$$

$$\widehat{PS}_{sys}(t+1) = fuzzy\left(\widehat{PS}_j(t+1)\right) \tag{13}$$

$$\widehat{R}_j(t+1) = \frac{\widehat{P}_j(t+1)}{P_{lj}} (j = 1, 2 \ldots, 13) \tag{14}$$

where $\widehat{PS}_{sys}(t+1)$ is the fuzzy index for system load flow security. $\widehat{PS}_j(t+1)$ shows the load security level of line j. $\widehat{R}_j(t+1)$ is the predictive active power rating ratio. $\widehat{P}_j(t+1)$ is the predicted active load flow of line j from CCN which is shown in Equation (2). P_{lj} is the active power rating.

4.4. Cellular Based Rule Base of HCRFS Block

The fuzzy implication of the first layer in the HCRFS was a multi-input–multi-output (MIMO) system:

Rule i: If a_1 is A_{1i} ... and a_r is A_{ri}, then b_1 is B_{1i} ... , and bn is B_{ni}.

where *i* is the number of rules, *r* is the input number and *n* is the output number.

The fuzzy implication of the second layer in the HCRFS follows multi-input–single -output (MISO) systems:

Rule j: if x_1 is A_{1j} ... , and x_r is A_{rj}, then y is C_j.

where *i* is the number of rules, *r* is the input number.

In the 12-bus power system, ideally to classify the power system states "Secure", "Alert", or "Emergency", it requires abundant combinations of input levels. Assume each of the 11 voltage inputs is classified by five levels (corresponding membership functions are NB, NM, Normal, PM, PB); this would mean 5^{11} = 48,828,125 voltage level combinations.

In order to overcome the rule explosion, an idea of the cellular-based rule base is inspired by the CCN structure. CCN is a cellular computational network consisting of a neural network in each cell that can be used to implement networked power systems. In the design of the voltage magnitude prediction layer, CCN was applied to implement the connection of the 11 buses. Each cell/bus of the CCN could be trained and used separately only considering the information of the nearest neighbors. Similarly, as the status of the system security had a tight relationship with contingency cases who may lead to voltage violation or overload, different types of contingencies were considered in developing rules process for the cellular concept based fuzzy system.

In the design of the cellular-based rule base, only the buses directedly connected to the contingency were considered. For example, if there was a line contingency between bus 1 and 2, only 2 primary buses (buses 1 and 2) were considered in the "IF" condition part, instead of all the inputs.

From Table 2, there are 13 outages. If the cellular-based fuzzy rules were applied for each outage using only 2 primary buses, the rule cases reduced to 5^2 = 25 voltage level combinations for each outage, and 13 outages meant $13 \times 25 = 325$ rules. Finally, it is more than $1 - (325 \div 48{,}828{,}125) = 1 - 0.00066\% = 99.99\%$ reduction in the number of rules.

5. Visualization

After the projection process, the buses and system security level were displayed geographically two-dimensionally using web-based computer language.

5.1. Bus Voltage and Line Load Flow Security Visualization

With the application of computer language, the security level of each element and the power system could be vividly displayed to the operation staff. As shown in Figure 1, all the predicted security information from the HCRFS block is displayed geographically two-dimensionally. The displayed information includes the voltage magnitude security level of each bus, the load flow security of each line, and the overall power system security level from the viewpoint of the whole power network.

The power system security states were characterized in three modes from the view of the power system operators, which were Secure, Alert, and Emergency:

- Secure: All the buses are in normal states, which means there is no alarm being presented and none of the contingency would cause overload or voltage violations;
- Alert: There is an alarm or contingency which needs the operator to pay attention;
- Emergency: It is indicated that a serious alarm appears, and the system is seriously insecure, or there is a contingency that may lead to system blackout which needs the operator to act immediately.

The voltage magnitude security of each bus was shown geographically above the position of each bus in the 12-bus power system as in below:

$$V_{security\ Level} = \begin{Bmatrix} Secure & Green\ Pie \\ Alert & Yellow\ Pie \\ Emergency & Red\ Pie \end{Bmatrix} \tag{15}$$

The positive "+" and negative "−" showed the voltage magnitude was above or below 1 pu. The diameter of the symbol increased with the rise of the security level to make it more noticeable.

The overload situation of each line was displayed in a circular arrow manner above that line using different colors.

$$P_{security\ Level} = \begin{Bmatrix} Secure & Green\ Circle \\ Alert & Yellow\ Circle \\ Emergency & Red\ Circle \end{Bmatrix} \tag{16}$$

5.2. Power System Security Level Visualization

The prediction of the power system security level from HCRFS was visualized in a meter to make it easy to read. Secure, Alert, and Emergency were labeled on the pane of the meter as a reference for the pointer to show the security level.

6. Results and Discussion

From Section 3, the voltage and active power flow prediction was carried out based on the two-layer CCN model. The power system steady-state security related to its robustness under the normal state, as well as to withstanding foreseeable contingencies without interruption to customer service. Thus, the CCN-based power system security analysis was done from two aspects: 1. Voltage and active power prediction with pseudorandom binary sequence (PRBS) signals on generators or load to simulate normal disturbance of the 12-bus power system; 2. voltage and active power prediction under single line outage to test the power system security in the event of unforeseen contingency. The details of the simulation cases are shown in Table 3. Case A is a training case with PRBS signals applied on generators G_2, G_3, G_4 to simulate the small noise and disturbance in the power system. The 0.5, 1, and 2 Hz PRBS signals were fluctuated positive and negative 15%, which were simulated voltage magnitude violations under the steady-state. Cases B to E were test cases.

Table 3. Simulation cases under different disturbances and outages.

Case No.	Disturbance Type
Case A	PRBS signals on G_2, G_3, G_4 (for batch training)
Case B	PRBS signals on loads of Bus_2, Bus_3, Bus_5, and $Line_{5\_4}$ outage
Case C	PRBS signals on load of Bus_2 and generator G_3
Case D	PRBS signals on loads of Bus_2, Bus_3, Bus_5, and $Line_{2\_5}$ tripped

In order to see the performance of ESN-based CCN, the MLP- and RNN-based CCN predictions [27,28] were applied in this paper for comparison. Different from the online learning of ESN, the weights of MLP/RNN-based CCN were generated from batch training using dynamic multi-swarm particle swarm optimization (DMSPSO) [33], and later applied to online prediction. Case A in Table 3 is the batch training data for MLP and RNN. In the batch training process, PRBS signals were applied on three generators to simulate the disturbance during actual system operation. The trained weights of MLP and RNN in each CCN cell were fixed and applied to online prediction in different scenarios. The parameter settings of the three kinds of neural networks are shown in Table 4. The stop iteration numbers for batch training were 3000, but for online training, the stop criteria was MSE $< 1 \times 10^{-2}$.

Table 4. Parameter settings for different types of neural networks.

Neural Networks Type	Training Type	Training Method	Max Iter.	Search Range	Group Num.	Particle Num.
MLP	Batch training	DMPSO	3000	[−1.5 1.5]	4	3
RNN	Batch training	DMPSO	3000	−[2 2]	5	3
ESN	Online Training	Line regression	MSE $< 1 \times 10^{-2}$	[−2 2]	-	-

6.1. Voltage Prediction with PRBS Signals on Generators (Case A)

In Figure 3, 1 s ahead voltage prediction results under PRBS signals on generators G_2, G_3, and G_4 can be seen. The solid blue line indicates PMU measurements from the 12-bus RTDS model (real data), the green dot-dash line is voltage predictions from MLP, the black dashed line shows results of RNN, and the red dotted line indicates voltage prediction of ESN. The panels in Figure 3 include voltage prediction results for three generator buses (Bus_{G2}, Bus_{G3}, and Bus_{G4}) and three load buses (Bus_{L4}, Bus_{L5}, Bus_{L6}). The oscillations on the red dotted line within 0 to 200 ms of Figure 3 come from the initiation process of ESN online training. From comparison, all three kinds of neural networks can follow the voltage change trend with a slight time shift.

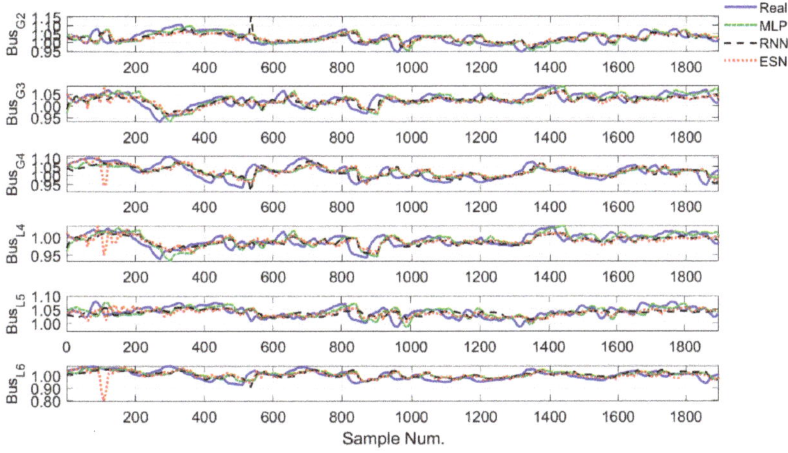

Figure 3. The 1 s voltage prediction near generator buses and load buses.

6.2. Voltage Prediction with PRBS Signals on Load Buses and Line Contingency (Case B)

Besides PRBS signals, transmission line contingency was also applied for several seconds and restored in this part for testing. Figure 4 shows voltage prediction results under PRBS signals on load buses (Bus_{L2}, Bus_{L3}, Bus_{L5}) and $Line_{5\_4}$ (which connected bus_4 and bus_5) outage. Because $Line_{5\_4}$ tripped, loads on bus_4 and bus_5 experienced slightly low voltage. As Bus_{L6} and Bus_{G4} were close to generators, the load on Bus_6 and generator on Bus_{G4} performed better with small oscillation when contingency occurred and disappeared. From Figure 4, the MLP- or RNN-based voltage prediction had a steady-state error, while the ESN prediction had the advantage of fast reaction and small overshoot when contingency occurred and was restored.

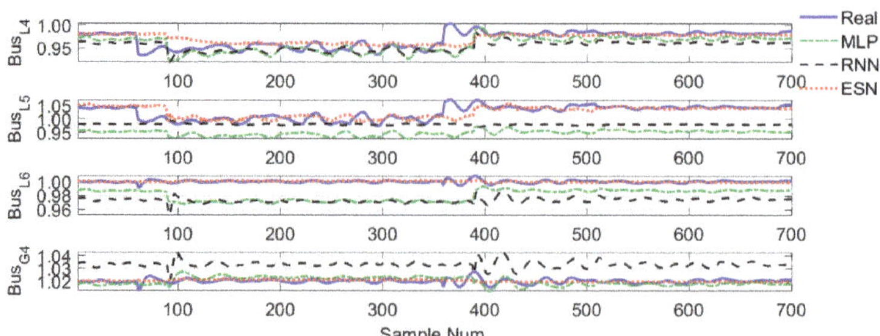

Figure 4. The 1 s voltage prediction on load buses and generator bus G_4 under pseudorandom binary sequence (PRBS) noise and $Line_{5\text{-}4}$ outage.

6.3. Load Flow Prediction with PRBS Signals on Generator and Load Buses(Case C)

A scheme with PRBS signals on load Bus_3 and generator G_2 was proposed here to see the performance of the active power prediction. In Figure 5, the performances of three kinds of neural networks were similar except for small differences. The MLP prediction performed better on peak values, while the online ESN method needed an adjusting time (simple 200 to simple 300) for initialization. The green dot-dash line in the $Line_{4\_3}$ panel looks like an average line in the horizontal direction. This phenomenon shows the MLP-based active power prediction cannot converge on Line $_{4\_3}$.

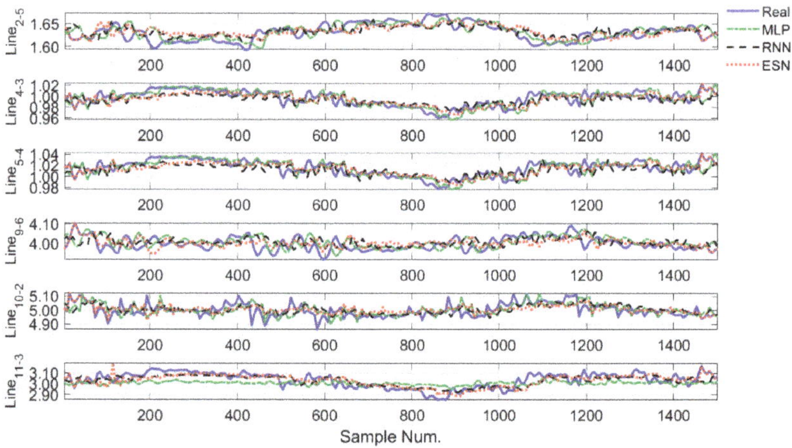

Figure 5. The 1 s load flow prediction of $Line_{2\text{-}5}$, $Line_{4\text{-}3}$, $Line_{5\text{-}4}$ (near load buses), $Line_{9\text{-}6}$, $Line_{10\text{-}2}$, and $Line_{11\text{-}3}$ (near gen. buses).

6.4. Load Flow Prediction with PRBS Signals on Load Buses and Line Contingency (Case D)

Figure 6 indicates the power flow change under load disturbance and $Line_{7\_8}$ break. The $Line_{1\_2}$ and $Line_{2\_5}$ undertook half of the loss led by the $Line_{7\_8}$ fault. The 1 s load flow forecast illustrated that the ESN method had excellent performance in line disconnection circumstances. From the comparison of all three methods, the online training ESN method overmatched the fixed weight RNN and MLP method as the online learning mechanism could update the weights of ESN in each cell timely, according to the situation change such as load noise of abrupt line off.

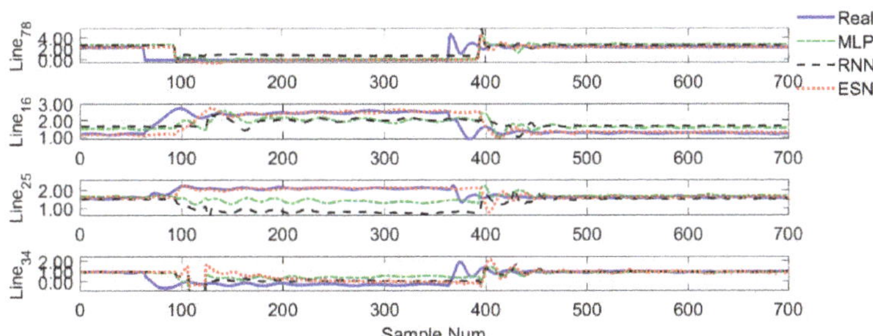

Figure 6. The 1 s ahead active power prediction on load Bus_3, Bus_4, Bus_5, and generator Bus_{G3} under PRBS signals on generators and $Line_{7\text{-}8}$ outage.

6.5. Performance Evaluation

As the performance of different neural networks was similar in Case A and Case C, the MSE errors between real values and predictions are shown in Table 5 for performance evaluation. It is reasonable that MLP/RNN performed better in Case A than ESN as Case A was a batch training case, while ESN needed initialization time for online training. But in Case C, ESN had better prediction results as the online training method could change the neural network weights with the situation change.

6.6. HCRFS based Power System Security Awareness

From Section 4, the two-layer HCRFS block could estimate the whole system security situation including component (bus and transmission line) security and system security. In the application of web-based computer visualization technology, the output voltage and load flow security level of the HCRFS could be vividly displayed to the operation staff in a visual manner. Figures 7–9 are the visualization results.

Figure 7 is the web-based graphic user interface under a normal operation state. The left graph is the component visualization (voltage security on each bus and load flow security on each line). The right meter is the corresponding system security display. From Figures 7–9, different colors indicate different security states of the system. It is defined in Equations (15) and (16) that green means the Secure state, orange shows the Alert state, while red indicates the Emergency state. The pie shapes represent the voltage security of each bus geographically. The circular arrows illustrate the load flow security level of each transmission line with various colors. Normally, the system is in the Secure state (Figure 7). The green pies on the buses and green circles on the lines show that the bus voltage and line load flow are all fluctuating within a secure range.

The security situation of the power system changes under disturbances, as shown in Figures 8 and 9. To highlight the main parts, Figures 8 and 9 remove style designs including title, background, and logo, zooming in on the geographic two-dimensional based component security and system security meter.

6.6.1. Bus Voltage and Line Load Flow Security Awareness under Small Disturbance

The visualization of each bus and transmission line under PRBS signals are shown in Figure 8a. Because of the disturbances (PRBS signals) on generators G_2, G_3, and G_4, there were some Alert states compared with the initial Secure states (Figure 7). In Figure 8, the yellow negative pies above the buses reveal that Bus_3, Bus_4, Bus_5, Bus_6, Bus_8, Bus_9, and Bus_{11} are slightly below the rated voltage. At the same time, yellow circles above the transmission $line_{1\_6}$, $line_{1\_7}$, $line_{7\_8}$, and $line_{8\_3}$ indicate slight overflow on those lines.

Table 5. Mean square error (MSE) between real values and predictions of different neural networks.

Neural Networks Type	Case A			Case C		
	Max. MSE	**Min. MSE**	**Mean MSE**	**Max. MSE**	**Min. MSE**	**Mean MSE**
MLP	9.5594×10^{-4}	8.7512×10^{-5}	5.2308×10^{-4}	0.0067	1.1533×10^{-4}	0.0015
RNN	0.0011	4.5032×10^{-5}	4.6935×10^{-4}	0.0047	1.1528×10^{-4}	0.0011
ESN	0.0012	4.7166×10^{-5}	5.2835×10^{-4}	0.005	8.7193×10^{-5}	9.9694×10^{-4}

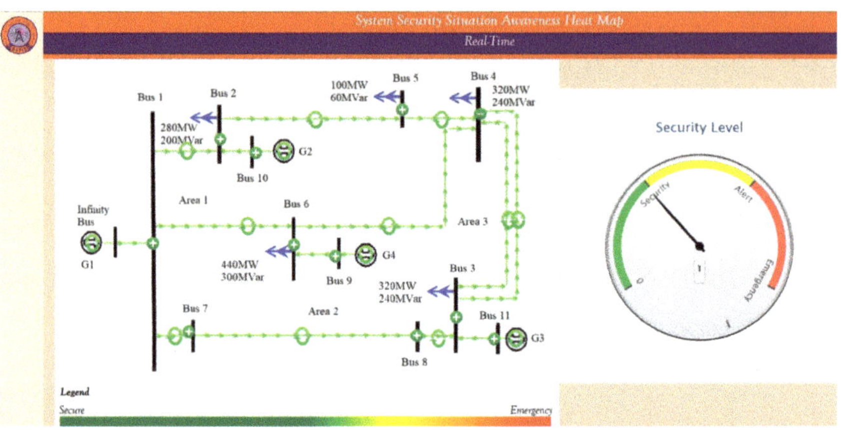

Figure 7. Graphic user interface of the situation awareness system (initial state).

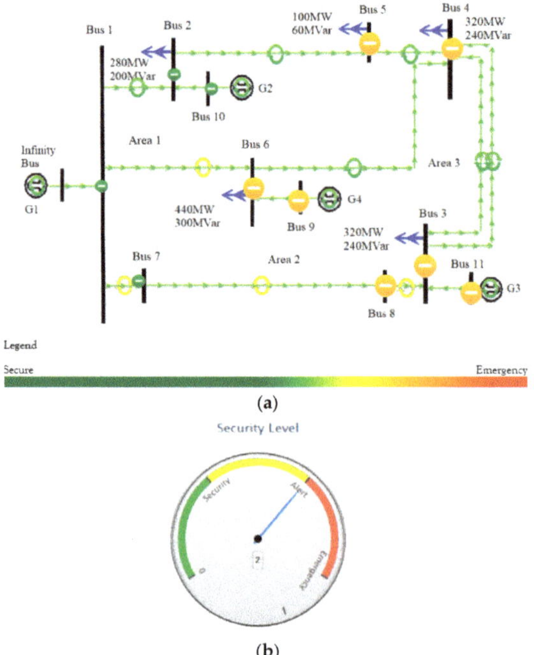

Figure 8. Graphic user interface under PRBS signals disturbance. (**a**) The components' security awareness of the 12-bus power system, (**b**) The system security state.

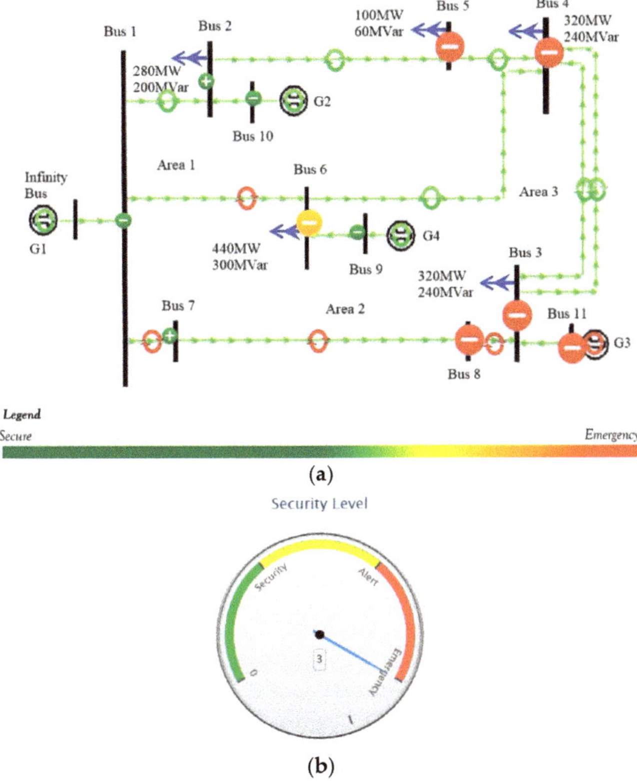

(a)

(b)

Figure 9. Graphic user interface state under PRBS signals and $line_{1\_2}$ contingency. (**a**) The components' security awareness of the 12-bus power system, (**b**) The system security state.

6.6.2. Bus Voltage and Line Load Flow Security Awareness under Small Disturbance and Line Contingency

Both PRBS signal and $Line_{1\_2}$ contingencies are applied in this simulation in Figure 9. Compared with the entire green in Figure 7, the 11 buses have 1 Alert state (yellow pie) and 5 Emergency states (red pie), while the 13 transmission lines have 5 Emergency states. Because $Line_{1\_2}$ is tripped, generator G_3 and transmission lines $Line_{1\_6}$, $Line_{1\_7}$, $Line_{8\_3}$, and $Line_{7\_8}$ are seriously overloaded (red circle). The five red pies show that because of the disturbance and $Line_{1\_2}$ contingency, the power system is instable, and Area 3 is in a serious sub-voltage situation.

6.6.3. Power System Security Level Visualization

The system security was defined in a meter considering all the buses' and transmission lines' secure state. The different states in the meters of Figures 7b, 8b and 9b were in accordance with the bus and line secure state from Figures 7a, 8a and 9a. The pointer in Figure 7b indicates that the system security level is Secure, which is consistent with the initial Secure state of the buses and transmission lines. Alert states on 7 buses and 4 lines in Figure 8a indicate that sub-voltage and overload appeared on the system. That is why the indicator points to Alert in Figure 8b at that time instant. The system security in Figure 9b is in Emergency level, corresponding to the 5 serious buses voltage violation and 5 severe lines' surcharge in Figure 9b.

7. Conclusions

This paper has presented how to implement power system security awareness using a CCN and HCRFS combined methodology. From the results presented, it can be seen that the ESN-based CCN bus voltage and line load flow prediction could estimate the state of the power system online better than MLP- and RNN-based CCNs, with a mean MSE lowered to 9.9694×10^{-4} under new events or contingency. The proposed HCRFS system reduces the number of rules in the rule-base dramatically by 99.99%. The two-dimensional visualizations could vividly display the bus voltage security levels, transmission line power flow security states, and the system security situation synchronously to the control room operator. The predicted security level could inform the system operator to react in advance to prevent a cascaded contingency and even a system blackout. Multiple results show that the proposed CCN and HCRFS combined visualization method could predict the security of the power system with acceptable accuracy under both small disturbance and line contingency. Future work includes: Improving the prediction accuracy, the dynamic security could be considered later, and parallel computing could be applied to improve the training efficiency. Furthermore, the proposed CCN and HCRFS combined system security level prediction and visualization technique can be applied to a 68-bus system to study the scalability of the proposed CCN- and HCRFS-based approach.

Author Contributions: All authors have contributed to the publication of this article. L.W. and G.K.V. worked together on the applications of cellular computational network (CCN) for the proposed study while first author was a visiting researcher at Clemson University in 2018 and 2019. L.W., G.K.V. and J.G. work together on proposed content of this paper; L.W. wrote the paper with guidance and advice from G.K.V. All authors have read and agreed to the published version of the manuscript.

Funding: This work was supported by # 1312260 and Duke Energy Distinguished Professorship Endowment, the National Natural Science Foundation of China (61803343), Key R\&D and Promotion Project of Henan Province (202102210096, 202102210296), and Key Projects of Higher Education of Henan Province (19A120012).

Institutional Review Board Statement: Not applicable.

Informed Consent Statement: Not applicable.

Data Availability Statement: Not applicable.

Acknowledgments: Lili Wu acknowledges the Real-Time Power and Intelligent Systems (RTPIS) Laboratory at Clemson University where she was a visiting researcher from January 2018 to December 2019. For the research study reported in this paper, Lili Wu acknowledges the contributions of RTPIS Lab and the research assistants including (i) Paranietharan Arunagirinathan for assisting with the power system model and its simulation on the real-time digital simulator and (ii) Iroshani Jayawardene for assisting with the visualization design and discussions on the cellular computational network.

Conflicts of Interest: The authors declare no conflict of interest.

References

1. Kundur, P.; Paserba, J.; Ajjarapu, V.; Andersson, G.; Bose, A.; Canizares, C.; Hatziargyriou, N.; Hill, D.; Stankovic, A.; Taylor, C.; et al. Definition and classification of power system stability. *IEEE Trans. Power Syst.* **2004**, *19*, 1387–1401. [CrossRef]
2. Debs, A.S.; Larson, R.E. A Dynamic Estimator for Tracking a Power System. *IEEE Trans. Power Appar. Syst.* **1970**, *89*, 1670–1678. [CrossRef]
3. Xiao, F.; Jiang, Z.-Q.; Ai, Q.; Hao, R. Situation awareness of power system based on static voltage security region. *J. Eng.* **2017**, *2017*, 2423–2427. [CrossRef]
4. Netto, N.A.R.L.; Borges, C.L.T. Enhancing the situational awareness of voltage security region via probabilistic reliability evaluation. *Int. Trans. Electr. Energy Syst.* **2020**, *30*, 1–15. [CrossRef]
5. Sun, C.; Liu, D.; Wang, Y. Operation Situational Awareness Based on Dynamic Power Flow for a Profound Analysis of Active Distribution Network. In Proceedings of the IET Conference Publications, Helsinki, Finland, 14–15 June 2016; pp. 2–5.
6. Kalyani, S.; Swarup, K.S. Particle swarm optimization based K-means clustering approach for security assessment in power systems. *Expert Syst. Appl.* **2011**, *38*, 10839–10846. [CrossRef]

7. Matos, M.A.; Hatziargyriou, N.D.; Lopes, J.A.P. Multicontingency steady state security evaluation using fuzzy clustering techniques. *IEEE Trans. Power Syst.* **2000**, *15*, 177–183. [CrossRef]
8. Krishnakumar, B.; Subaashini, M.; Kumar, E.G.; Arthi, R. Power flow based contingency analysis using fuzzy logic. *Procedia Eng.* **2012**, *38*, 3603–3613. [CrossRef]
9. Ozdemir, A.; Singh, C. Fuzzy Logic Based MW Contingency Ranking against Masking Problem. In Proceedings of the 2001 IEEE Power Engineering Society Winter Meeting, Columbus, OH, USA, 28 January–1 February 2001; pp. 504–509.
10. Marannino, P.; Berizzi, A.; Merlo, M.; Demartini, G. A rule-based fuzzy logic approach for the voltage collapse risk classification. *Proc. IEEE Power Eng. Soc. Transm. Distrib. Conf.* **2002**, *2*, 876–881. [CrossRef]
11. Halilčević, S.S.; Gubina, F.; Gubina, A.F. The uniform fuzzy index of power system security Suad. *Eur. Trans. Electr. Power* **2010**, *20*, 785–799. [CrossRef]
12. Halilčević, S.S.; Gubina, F.; Gubina, A.F. The composite fuzzy reliability index of power systems. *Eng. Appl. Artif. Intell.* **2011**, *24*, 1026–1034. [CrossRef]
13. Kalyani, S.; Swarup, K. Supervised fuzzy C-means clustering technique for security assessment and classification in power systems. *Int. J. Eng. Sci. Technol.* **2010**, *2*, 175–185. [CrossRef]
14. Zhao, J.; Zhou, Y.; Shuo, L. A Situation Awareness Model of System Survivability Based on Variable Fuzzy Set. *TELKOMNIKA Indones. J. Electr. Eng.* **2012**, *10*, 2239–2246. [CrossRef]
15. Fan, S.; Lin, J.; Zhang, T.; Dong, G.; Liu, X. Temporal and Spatial Distribution of Power System Voltage based on Generalized Regression Neural Network. *ISPEC* **2019**, 1891–1896. [CrossRef]
16. Diao, R.; Vittal, V.; Logic, N. Design of a real-time security assessment tool for situational awareness enhancement in modern power systems. *IEEE Trans. Power Syst.* **2010**, *25*, 957–965. [CrossRef]
17. Özdemir, A.; Lim, J.Y.; Singh, C. Post-outage reactive power flow calculations by genetic algorithms: Constrained optimization approach. *IEEE Trans. Power Syst.* **2005**, *20*, 1266–1272. [CrossRef]
18. Javan, D.S.; Mashhadi, H.R.; Rouhani, M. A fast static security assessment method based on radial basis function neural networks using enhanced clustering. *Int. J. Electr. Power Energy Syst.* **2013**, *44*, 988–996. [CrossRef]
19. Sidhu, T.S. Contingency screening for steady-state security analysis by using FFT and artificial neural networks. *IEEE Trans. Power Syst.* **2000**, *15*, 421–426. [CrossRef]
20. Luitel, B.; Venayagamoorthy, G.K. Cellular computational networks-A scalable architecture for learning the dynamics of large networked systems. *Neural Netw.* **2014**, *50*, 120–123. [CrossRef]
21. Wang, S.; Gao, W.; Meliopoulos, A.P.S. An alternative method for power system dynamic state estimation based on unscented transform. *IEEE Trans. Power Syst.* **2012**, *27*, 942–950. [CrossRef]
22. Hasala Dharmawardena, G.K.V. Cellular Computational Network for Distributed Power Flow Inferencing in Electric Distribution Systems. In Proceedings of the IJCNN, Budapest, Hungary, 14–19 July 2019; pp. 1–8.
23. Rahman, M.A.; Venayagamoorthy, G.K. Scalable Cellular Computational Network Based WLS State Estimator for Power Systems. In Proceedings of the 2015 Clemson University Power Systems Conference, Clemson, SC, USA, 10–13 March 2015.
24. Rahman, M.A.; Venayagamoorthy, G.K. Power System Distributed Dynamic State Prediction. In Proceedings of the 2016 IEEE Symposium Series on Computational Intelligence, Athens, Greece, 6–9 December 2016; pp. 1–6.
25. Balasubramaniam, K.; Luitel, B.; Venayagamoorthy, G.K. A scalable wide area monitoring system using cellular neural networks. *Proc. Int. Jt. Conf. Neural Netw.* **2012**, 10–15. [CrossRef]
26. Balasubramaniam, K.; Venayagamoorthy, G.K.; Watson, N. Cellular neural network based situational awareness system for power grids. *Proc. Int. Jt. Conf. Neural Netw.* **2013**. [CrossRef]
27. Wu, L.; Venayagamoorthy, G.K.; Harley, R.G.; Gao, J. Cellular Computational Networks Based Voltage Contingency Ranking Regarding Power System Security. In Proceedings of the Clemson University Power Systems Conference, Charleston, SC, USA, 4–7 September 2018.
28. Wu, L.; Venayagamoorthy, G.K.; Gao, J. Cellular Computational Networks for Distributed Prediction of Active Power Flow in Power Systems under Contingency. In Proceedings of the Proceedings of 2019 IEEE PES Innovative Smart Grid Technologies Europe, Bucharest, Romania, 29 September–2 October 2019; pp. 1–4.
29. Shan, J. A platform for validation of FACTS models. *IEEE Trans. POWER Deliv.* **2006**, *21*, 484–491. [CrossRef]
30. Jaeger, H. The "echo state" approach to analysing and training recurrent neural networks–with an Erratum note. *GMD* **2001**, *148*, 13.
31. IEEE. *IEEE Technical Report IEEE/PES Voltage Stability of Power Systems: Concepts, Analytical Tools, and Industry Experience*; IEEE: New York, NY, USA, 18 August 1990.
32. Stott, B.; Alsac, O.; Monticelli, A.J. Security Analysis and Optimization. *Proc. IEEE* **1987**, *75*, 1623–1644. [CrossRef]
33. Liang, J.J.; Suganthan, P.N. Dynamic Multi-Swarm Particle Swarm Optimizer. In Proceedings of the IEEE on Swarm Intelligence Symposium, Pasadena, CA, USA, 8–10 June 2005; pp. 124–129.

Article

Estimation of Modal Parameters for Inter-Area Oscillations Analysis by a Machine Learning Approach with Offline Training

Carlo Olivieri [1], Francesco de Paulis [1], Antonio Orlandi [1,*], Cosimo Pisani [2], Giorgio Giannuzzi [2], Roberto Salvati [2] and Roberto Zaottini [2]

1 UAq EMC Laboratory, Department of Industrial and Information Engineering and Economics, University of L'Aquila, 67100 L'Aquila, Italy; carlo.olivieri@univaq.it (C.O.); francesco.depaulis@univaq.it (F.d.P.)

2 TERNA S.p.A., V.le Egidio Galbani, 70, 00156 Rome, Italy; cosimo.pisani@terna.it (C.P.); giorgio.giannuzzi@terna.it (G.G.); roberto.salvati@terna.it (R.S.); roberto.zaottini@terna.it (R.Z.)

* Correspondence: antonio.orlandi@univaq.it

Received: 29 October 2020; Accepted: 1 December 2020; Published: 4 December 2020

Abstract: An accurate monitoring of power system behavior is a hot-topic for modern grid operation. Low-frequency oscillations (LFO), such as inter-area electromechanical oscillations, are detrimental phenomena impairing the development of the grid itself and also the integration of renewable sources. An interesting countermeasure to prevent the occurrence of such oscillations is to continuously identify their characteristic electromechanical mode parameters, possibly realizing an online monitoring system. In this paper an attempt to develop an online modal parameters identification system is done using machine learning techniques. An approach based on the development of a proper artificial neural network exploiting the frequency measurements coming from actual PMU devices is presented. The specifically developed offline training stage is fully detailed. The output results from the dynamic mode decomposition method are considered as reference in order to validate the machine learning approach. Some results are presented in order to validate the effectiveness of the proposed approach on data coming from recordings of real grid events. The main key points affecting the performance of the proposed technique are discussed by means of proper validation scenarios. This contribution is the first step of a more extended project whose final aim is the development of an artificial neural networks (ANN) architecture able to predict the system behavior (in a given time span) in terms of LFO modal parameters, and to classify the contingencies/disturbances based on an online training that has memory of the passed training samples.

Keywords: inter-area oscillations; modal analysis; reduced order modeling; dynamic mode decomposition; machine learning; artificial neural networks

1. Introduction

It is a matter of fact that the actual ever-demanding environmental policies are forcing the worldwide power grids to integrate a rising amount of renewable sources, thus leading the grids themselves to become more and more interconnected, complex, and also prone to be stressed in their ordinary functioning.

Modern power systems have the fundamental need to deliver the largest electrical power over long distances. However, such systems should also be able to take into account the presence of renewable sources, characterized by very low inertia, which constitutes an impairment in terms of system's stability since they introduce rapidly changing electrical dynamics.

Under this scenario, the power grid utilities are often operated at the edge of their capacity and stability limits, where the possible presence of small disturbances due to switching or line operation events can negatively affect the reliability of the entire power system. Electromechanical inter-area oscillations constitute an inherent dynamic property of electric power system [1]. Such oscillations cannot be entirely suppressed and some of them are unavoidably excited by the mentioned disturbances. This kind of oscillatory phenomena is usually characterized by modal frequencies (named "inter-area" oscillations or modes) in a range between 0.1 and 0.8 Hz while other low-frequency oscillation (LFO) phenomena are characterized by relevant modes in a frequency range between 0.8 Hz and 2 Hz. These last oscillations are the so-called "local-area modes." The LFOs are mainly due to the swinging of one or more units at a specific generating station, or few generating stations that are geographically close, with respect to the rest of the power system; they can be mitigated through the use of local stabilization agents such as power system stabilizers (PSS), automatic voltage regulators (AVRs), FACTS devices, etc., [2]. Inter-area oscillations, instead, are associated to the swinging of many generating units or power plants belonging to one or more zones of the system that are geographically far from each other, against the units in the remaining parts of the power system. Thus, because of their global nature and to the very low frequencies involved, they are hard to be controlled and hence damped. It must be said that there is no strong definition of well-damped low-frequency oscillations, but a generally accepted rule of thumb defines an oscillation as sufficiently damped if the damping ratio is above the range 3–5% [3]. Under-damped or undamped oscillations can lead to large power swings or also to the tripping of protection relays and subsequent disconnection of parts of the grid and/or loads.

Therefore, by considering the issues induced by LFO, and in particular by the electromechanical inter-area oscillations, the possibility to monitor the power system in real-time in order to guarantee safe and reliable grid operation is a relevant need. Nowadays, the role of wide-area measurements systems (WAMS) are getting more and more attention, for what concerns the use of phasor measurement units (PMUs), for identifying hazardous LFO modes because of inter-area oscillations. Identification of such LFO is a big challenge for the network operators. The identification of LFO modes has been widely studied in the past as the problem of identifying the characteristic parameters of exponentially damped sinusoids (EDSs) [4,5]. Many techniques have been proposed for this scope in several application fields [6–8], leading to promising results.

Even though the topic has been extensively studied, this problem still receives very wide attention [9–12], especially in the field of power systems [13,14]. The identification techniques are mainly classified as model-based techniques and measurement-based techniques or, in some cases, as parametric and non-parametric methods. Model-based techniques are based on the use of a proper system modeling of the power grid, and the subsequent extraction of the mode parameters by eigenvalues decomposition (EVD) analysis. Generally, they can achieve a relatively high accuracy; however, they are not completely suitable for online monitoring purposes because of some limitations in terms of computational burden and inherent uncertainties in system modeling. Measurement-based techniques (also referenced as "mode meters") seems to be more appealing in order to realize a real-time LFO monitoring system, since they are based only on proper signal processing methods, thus they can be considered to be model-free and inherently data-driven. The set of available measurement-based methods is quite large since there exists a certain number of basic estimation algorithms that have been further extended in order to overcome the specific weakness.

Among this huge set of techniques, the most popular methods can be considered those based on Fourier transform [15], Prony algorithm [16], Tufts–Kumaresan algorithm [17], Kalman filtering [18], wavelet analysis [19], Hankel singular value decomposition-variable projection method [20] and techniques based on Hilbert–Huang transform [21] and its related refinements based on empirical mode decomposition [22,23]. A set of recent works have focused on such kind of techniques, by exploring extensively their capabilities and drawbacks [23–25].

However, it seems that, as of today, the major good point of measurement-based methods is that they are inherently data-driven. In effect, some works have already explored the possibility to use these techniques in the context of a wider and more complex machine learning framework based on the use of artificial neural networks (ANNs), such as [26–30], either for LFO identification or EDSs characterization. In the above mentioned papers, the measurement-based estimation techniques are combined with ANN-based methods in order to identify the LFO mode parameters in real-time, or to extract other useful power grid information (like the system operating conditions or the generator coherency). In other applications, the data extracted from the ANN are also used to control and to damp the LFO phenomena [28,31]. It must be remarked that a key point inside the listed approaches is the need to operate a dimensionality reduction of the huge amount of data coming from PMUs. This is due to the fact that the use of machine learning techniques for modal analysis requires an offline training of the ANNs that depends, in terms of time required, on the size of the network and the size of the input data. Plenty of dimensionality reduction techniques are present in literature that can be suitable for this scope. Some very popular are the principal component analysis (PCA) [32], the independent component analysis (ICA), the dynamic mode decomposition (DMD) [33,34], and their further extensions and modifications.

This work is part of a global on-going research project whose final aim is the development of an ANN architecture able to predict and classify LFO phenomena based on an online training approach that keeps memory of the passed training samples. The present contribution focuses on the development of the ANN architectures and their validation by an offline training. The online strategy and associated algorithm development is still under investigation.

Differently from other recent approaches, in this paper an ANN-based strategy for the online monitoring of inter-area LFO modal parameters is presented. The DMD technique is explored and used as a dimensionality reduction method. Section 2 summarizes the relevant features of the DMD technique applied to the PMUs measured data. In Section 3 the architecture of the proposed ANN and the issues related to its training are discussed. Section 4 is devoted to the discussion of the numerical results and their comparison with some measurement in order to validate the proposed approach. Finally, in Section 5, some conclusions are drawn by discussing the advantages and limitations of the proposed method and possible research directions.

2. Modal Decomposition of Frequency Oscillations in Electric Power Systems

As already stated, the identification of LFO phenomena in power grids has been performed in previous research works following several different valid approaches. A mainstream class of techniques accounts for the execution of a modal analysis of the data measured and collected by the PMUs (i.e., as [24,25]). In this works the instantaneous frequency measurements coming from some PMUs are processed using some modal estimation techniques in order to extract the characteristic parameters of the main electromechanical modes, such as frequency, damping ratio, and amplitude. With this information, it is possible to identify the characteristics of a certain number of modes and hence also the dominant one. As a subsequent identification step, it is also possible to verify the presence of a hazardous inter-area oscillation by looking at the frequency range of the dominant mode and the values of its damping ratio.

For the purposes of this paper, we use as modal estimation technique, the DMD method, since it is capable of both identifying the modal parameters of LFO and to operate a dimensionality reduction, thus enabling an efficient design and training of the ANN-based LFO identification strategy.

The dynamic mode decomposition method was first introduced by Schmid [33] as a numerical procedure capable to extract dynamical features from flow data. It has been later enhanced and refined [34–36] in order to be used as a modal analysis technique capable to extract the modal parameters of EDSs.

The DMD theory is based on the collection of input data as a snapshot sequence of the following form:

$$X_1^N = \{x_1, x_2, \cdots, x_N\} \tag{1}$$

where x_i is the i-th snapshot and X_1^N is a data matrix whose columns represent the collection of the different snapshots, from the first up to the N-th. If each snapshot x_i is composed of M spatial samples, X_1^N is a M-by-N matrix.

The DMD method [33] is essentially founded on the assumption that the snapshots are related to each other via a linear mapping, defining a linear dynamical system as in (2).

$$x_{i+1} = Ax_i \tag{2}$$

Equation (2) is supposed to be approximately invariant during the time period between the two snapshots. Based on this, any collection of snapshots X_1^N can be split into two subsets, $X_1^{N-1} = \{x_1, x_2, \cdots, x_{N-1}\}$ and $X_2^N = \{x_2, x_3, \cdots, x_N\}$ for which the following relationship holds:

$$X_2^N = AX_1^{N-1} + r \tag{3}$$

In (3) the term r denotes the vector of residuals accounting for the dynamic behaviors that cannot be completely described by the linear mapping. The eigenvalues of matrix A are referred to as "DMD eigenvalues" and the eigenvectors of A as "DMD modes."

There are different algorithms suitable to implement the DMD method. In this work the one based on the singular value decomposition has been selected based on a similarity transformation and an eigenvalue decomposition (known as "DMD with SVD approach"). The principal steps to be followed in this case can be resumed according to [35]:

(1) Arrange the input data snapshots X_1^N into two time series X_1^{N-1} and X_2^N
(2) Compute the SVD of X_1^{N-1} as $X_1^{N-1} = U\Sigma W^T$
(3) Build the matrix $\widetilde{S} = U^T X_2^N W\Sigma^{-1}$ and compute its eigenvalues λ_i and eigenvectors v_i

The i-th DMD eigenvalue is λ_i and the associated DMD mode is $U\,v_i$. The DMD eigenvalues can be used to extract the modal parameters of the EDSs oscillatory behavior embedded inside the input data, enabling us to identify the nature of electromechanical LFOs based on a collection of measured data from PMUs.

A very interesting feature of the DMD estimation technique is that the dimension of the matrix $\widetilde{S}$ can be much lower than the original dynamic matrix A. Thus, even though the spatial sampling can be done on many PMU locations (e.g., M locations), the number of estimated modes can be much lower. This is a very important key point for the proposed approach because it allows to associate, for any arbitrary number of PMU measurement locations, a limited number of estimated electromechanical modes obtaining an efficient dimensionality reduction similar to what can be obtained by PCA-based techniques [26].

In all the cases analyzed in this paper, the input data to be fed to the DMD modal decomposition algorithm are constituted by a collection of measurements of the grid frequency taken by the Italian Transmission System Operator (TSO) TERNA at various PMU locations. The output of the DMD estimation method is constituted by the modal parameters of the LFO detected in the power grid.

It is worthy to note that, in the proposed approach, the extraction of the LFO modal parameters through the DMD estimation acts as a preliminary and necessary step to obtain a correct and efficient training set of data required by the ANN-based approach, as discussed in the next section.

3. Prediction of Low-Frequency Oscillations Modal Parameters by an Artificial Neural Network

The proposed ANN-based method to estimate the LFO modal parameters consists in the proper designing and training of an artificial neural network in order to be able to recognize the modal parameters associated to the LFO. This task is accomplished by considering the frequency measurements collected on the grid by a certain set of PMU devices. It must be remarked that the number of PMU measurement locations can be chosen arbitrarily; however, it must be set as a fixed parameter at the beginning of the procedure.

At a high-level description, as depicted in Figure 1, the raw frequency data coming from the PMU measurement system over a certain number of grid nodes should be preliminarily pre-processed in order to:

- Filter out the noise coming from the PMU measurements; this can be done with a properly tuned digital filter (for instance a classical low-pass FIR filter or a Hilbert filter);
- Detrend the data;
- Divide the overall measured input data stream in a certain number of data frames (i.e., "data windows") suitable to be used for the training stage of the ANN and to feed the neural network during the online LFO modal parameters identification process.

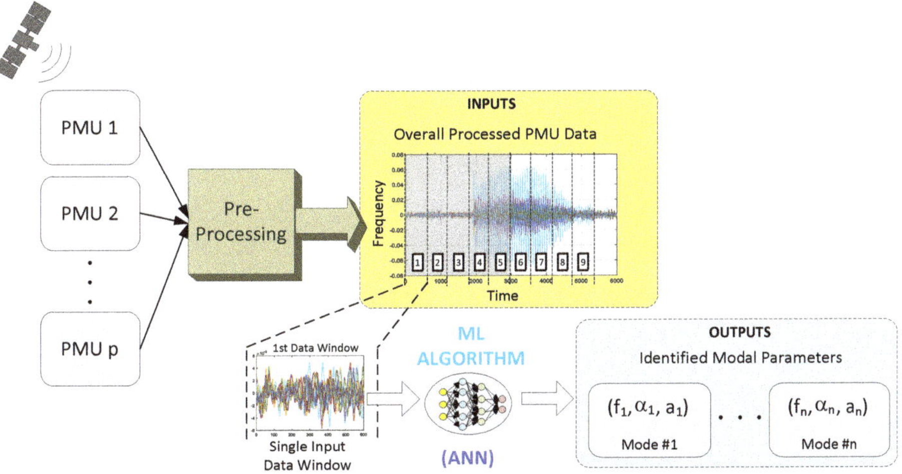

Figure 1. High-level schematization of the proposed low-frequency oscillations (LFO) modal parameters estimation technique.

After the pre-processing stage, the filtered data are given as the input to the ANN by using the sequencing defined by the data windows, such that the ANN is fed by using a sliding window mechanism.

For each data window gathered from the input data stream the ANN provides, as output, the estimated values of the LFO modal parameters; the parameters considered in this work are frequency (f_i), damping ratio (α_i), and amplitude (a_i) for each i-th electromechanical mode identified.

As illustrated in Figure 1, the ANN will provide, for each data window defined by the PMU and presented as its input, a triplet of values for each identified mode. As already mentioned, the number of modes to be identified can be selected arbitrarily by the user but, once it has been chosen, it has to be maintained as a constant parameter all along the calculations. In this paper, the number of modes of interest, based on the heuristic TSO experience, is equal to four.

3.1. Architecture of the Artificial Neural Networks

Among the many types of available artificial neural network architectures, the present study has focused its attention on the regression/estimation capabilities offered by one of the simplest types of ANN: the feed forward (FF) architectures. The reason for this choice is based on the final aim of the overall research project (not yet addressed in the present work): the ANN should be fed in real-time by the data coming from the PMUs and its output (the estimation of the LFO parameters) should be obtained within the shortest processing time. With the increasing of the complexity of the ANN structure, the training stage becomes more and more time-consuming, preventing the development of a machine learning framework capable to re-train the employed ANNs in a sufficiently short time period. Furthermore, it is known that the increase in complexity of a ANN is not a sufficient condition for an increase of its accuracy. On the contrary, there is the actual possibility to obtain the negative effect of overfitting, which would play in the present applications a fairly detrimental role.

For the actual application to LFO, two main classes of ANN architectures are considered, as depicted in Figure 2: the feed forward (FFNN) class and the cascade feed forward (CFNN) class.

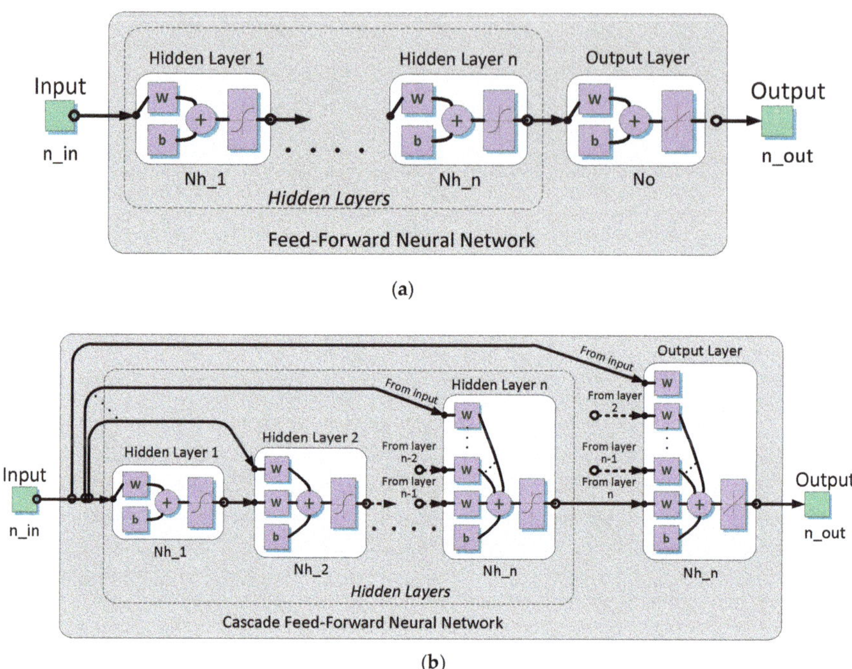

Figure 2. Proposed artificial neural networks (ANN) architectures: (**a**) feed forward and (**b**) cascade feed forward classes.

The motivation to explore two different kinds of ANN architectures is linked with the fact that, even though the FFNN class is already acceptable in order to get a baseline estimation performance, the CFNN class shows a faster learning rate. Therefore, it can be usefully compared with the FFNN class, in particular in terms of estimation accuracy. Table 1 summarizes the main design parameters (number of hidden layers, number of neurons for each layer, etc.,) for the proposed configurations that deserve to be investigated. Their values come from the experience and from a campaign of tests that are not described herein.

Table 1. ANN configurations and architectural parameters.

Configuration ID	Nr. of Hidden Layers	Nr. of Neurons for Each Layer	Hidden Layers Transfer Functions	Nr. of Neurons in the Output Layer	Output Layer Transfer Function
FFNN#1	2	10	TanSig	4	Linear
FFNN#2	3	10	TanSig	4	Linear
CFFNN	2	10	TanSig	4	Linear

3.2. Training of the Artificial Neural Networks

As appropriate for supervised learning architectures, the developed ANN must be trained to provide suitable regression (i.e., in this case estimation) features. The basic scheme reported in

Figure 3 illustrates how both types of ANN should be trained to be able to correctly identify the modal parameters of the LFO contained in the PMU measured data.

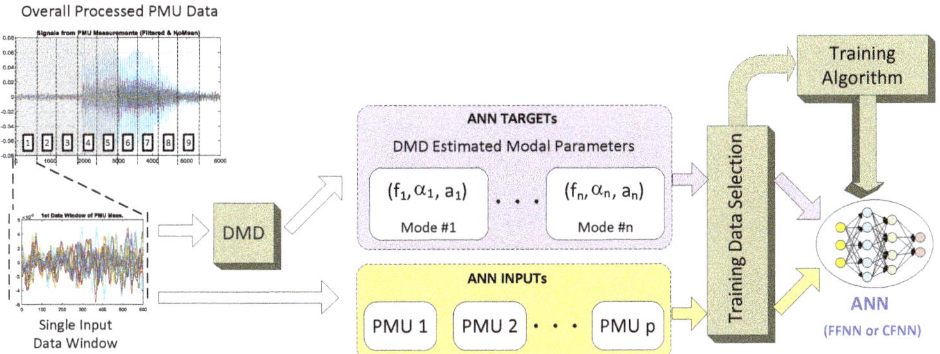

Figure 3. Overall scheme for the training stage of the ANNs.

The development of the input/target training set pair for the ANN starts from the PMU dataset that is suitably divided based on the mentioned sliding windows.

The input training set are the pre-processed PMU data. The corresponding target set are the output of the DMD procedure when its input are the pre-processed PMU data.

Once both the ANN target values and inputs have been obtained, the ANN training algorithm can be started. However, for this specific application, a point of attention should be further considered: since the input data originates from real PMU measurements, it happens sometimes that the gathered values can assume special non-numeric values (e.g., NaN values) because of events directly linked to the data acquisition system. In addition, NaN values can be also be numerically generated in our scenario when applying the DMD procedure, specifically when the DMD estimation algorithm finds some modes with frequencies lying outside a predefined frequency range.

As consequence of the presence on the NaN values it is mandatory, for the development of the present ANN-based LFO analysis technique, the proper handling of their occurrence in order to prevent that their presence alters the significance of the training set and could have a negative impact on the ANN-based performance. For this reason, during the training stage of the ANNs (both FFNN and CFFNN), a training data selection step is implemented. The NaN values are handled in the following manner:

- When the NaN value comes from the original PMU input data (i.e., it is inside the ANN inputs), the related data sample is excluded from the sliding window to which it belongs;
- When the NaN values come from the DMD estimation procedure (i.e., it is inside the ANN targets), the related target vector is treated as a "don't care" target, meaning that the network performance function is not updated during the training process for that specific target value.

The overall available PMU input data set is segmented in two parts: one used for training and one used as input for the estimation of the LFO ringing parameters. In turn, as usual for the supervised learning schemes, the training part is subdivided as 70% of samples used for training, 15% of them used for testing, and the remaining 15% used for validation.

4. Results and Validation

In this section some results obtained by using the proposed strategy of an ANN-based approach are presented in order to identify the LFO parameters characterizing oscillatory phenomena occurring in real power grid operation.

4.1. Validation Scenarios, Data Origin, and Details of the Considered Datasets

The overall data exploited to validate the effectiveness of the ANN approach for the estimation of the LFO modal parameters consists of three different datasets coming from real PMU measurements. Such data are collected by the Italian TSO TERNA from different electrical substations, and under different grid operating conditions. The three datasets can be described in the following way:

- Dataset 1 (in the following DS1): it contains the data coming from the measurements of an LFO oscillatory event (time range: 10 min), taken at 22 PMU measurement locations;
- Dataset 2 (in the following DS2): it contains the data coming from the measurements of a second LFO oscillatory event (time range: 20 min), taken at 30 PMU measurement locations;
- Dataset 3 (in the following DS3): it contains the measurement data collected from a 24 h recording of rated grid operation taken at 18 PMU locations.

All the data recordings have been obtained by using a sampling time of 100 ms and the corresponding samples have been collected into proper data files. According to what is already mentioned in Section 3, the datasets have been arranged for a proper windowing. Each data window is extracted from the original dataset considering a proper data frame length (this quantity is indicated as T_{DF} in the following), such that each dataset is composed of a different number of data windows.

A high-level outlook of the three datasets employed for this study is given in Figure 4. The first two datasets (Figure 4a,b) are characterized by ringing phenomena of the frequency measurements, especially at certain PMU locations. The third dataset instead is similar to the recording of nominal (also named as "ambient") conditions where some spurious frequency deviations occur from time to time.

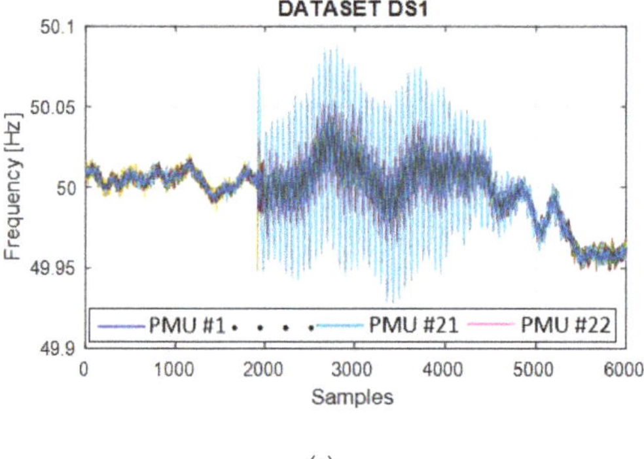

(a)

Figure 4. *Cont.*

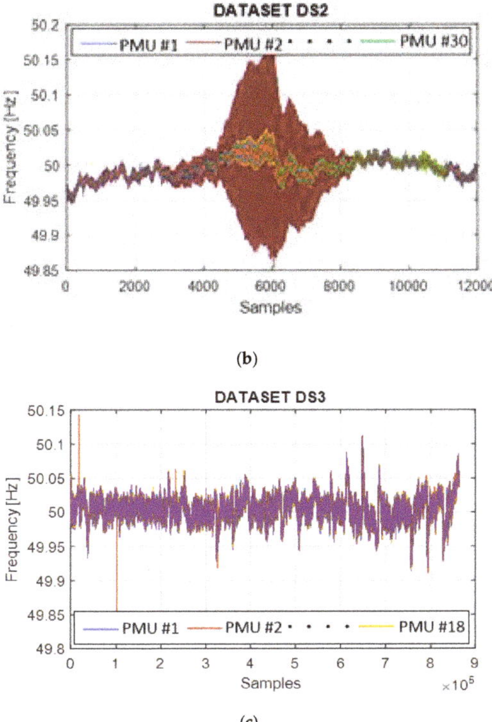

(**b**)

(**c**)

Figure 4. Overview of the data contained in the considered three datasets: (**a**) DS1: oscillatory event #1, (**b**) DS2: oscillatory event #2, (**c**) DS3: 24 h frequency recordings from nominal or "ambient" grid operation.

The duration of the data frame that is initially considered is equal to 20 s; this is the value that was found more relevant for this kind of analysis, based on the on-field experience acquired by the TSO.

Furthermore, each input data stream from the three datasets is pre-processed with a Hilbert filter having the parameters reported in Table 2, according to what is already mentioned in Section 3. The parameters listed in Table 2 specify the order of the filter, N, and the frequency edges of the transition bands of the filter, given by the parameters: f_1, f_2, f_3, f_4. The order of the filter is chosen depending on the length of the data frame.

Table 2. Parameters of the considered Hilbert filter.

Parameter	Value	Units
N	64	-
f_1	0.05	Hz
f_2	0.1	Hz
f_3	0.5	Hz
f_4	0.55	Hz

In order to assess the results obtained with the proposed ANN-based approach, a proper set of validation scenarios are proposed, where either the three considered datasets or the different architectures of the developed ANNs are systematically explored. An overview of these validation scenarios is reported in Table 3, where validation scenario VS#1 is considered for the assessment of the performance obtained by the ANN-based approach among the three different datasets. The other two validation scenarios VS#2 and VS#3 are considered to assess the performances of different ANN architectures over the same dataset, and to check the impact of the data frame length, respectively.

Table 3. Validation scenarios: datasets and ANN configurations.

Validation Scenario	Dataset ID	Network Configuration	Dataframe Length (T_{DF}) [s]	# of Windows
VS#1	DS1	FFNN#2	20	29
	DS2	FFNN#2	20	60
	DS3	FFNN#2	20	4321
VS#2	DS1	FFNN#1	20	29
	DS1	FFNN#2	20	29
	DS1	CFFNN	20	29
VS#3	DS1	CFFNN	30	19 (*)
	DS1	CFFNN	60	9 (*)

(*): Values subject to the effect of rounding on the number of windows considered.

4.2. Results of the LFO Modal Analysis and Impact of the Network Structure

As a first step, the capability of the ANN-based approach is evaluated to accurately extract the LFO parameters. Validation scenario VS#1 is considered for this scope. Figure 5 reports the comparison between the LFO parameters estimated by the ANN-based approach with respect to those obtained by applying the DMD method, as the reference technique. The plots are limited to the first two modes for the sake of simplicity, since they are the most relevant in terms of energy values.

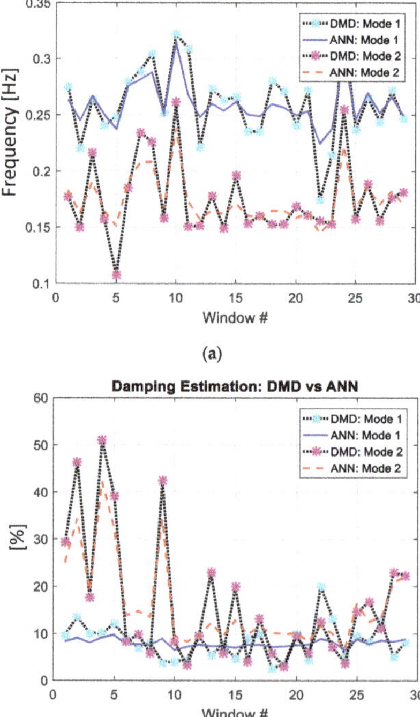

(a)

(b)

Figure 5. *Cont.*

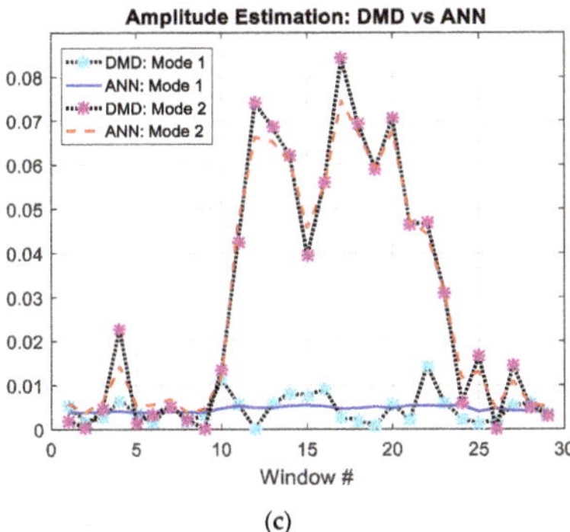

(c)

Figure 5. Comparison of the estimates produced by dynamic mode decomposition (DMD) versus the estimates produced by the ANN approach for the case of Dataset DS1: (**a**) frequency; (**b**) damping, (**c**) amplitude.

The LFO parameters estimated by both the DMD method and the ANN-based method are in good agreement, as can be seen from Figure 5. For the specific test case of DS#1 in Figure 5c, it is clear that the ANN is able to identify the principal contribution to the LFO phenomenon occurring in the grid, since it is in good agreement with the identification performed by the DMD technique. From Figure 5c, the principal mode is detected as the mode #2 (i.e., the one characterized by the greatest value of mode amplitude). Figure 5a shows as both the ANN and the DMD evaluate the modal frequency of the principal mode (mode #2) at around 0.15 Hz. This value is in agreement with what is already known from the theory about inter-area electromechanical oscillations, and from the consolidated know-how of the TSO for that specific grid event.

The ANN LFO parameter estimation accuracy continued within the validation scenario VS#1; Figure 6 reports the analysis of the parameters for the grid event captured in dataset DS2.

The proposed ANN-based approach is able to provide estimates of the LFO parameters having a good degree of confidence, if compared to those provided by the DMD method. This is confirmed especially if we look at the evaluations provided by the ANN for the inputs lying outside its training set, thus the estimates provided from the second half of each DS (specifically from window 30 up to window 60 in all the cases of Figure 6).

In this second test case the estimated mode amplitudes, provided by the ANN-based approach, indicate that mode #1 is the one contributing more to the LFO, which is an inter-area oscillation characterized by a frequency value around 0.3 Hz. Also, in this case the frequency value agrees with the theoretical background and with the data experienced by the TSO about this kind of inter-area oscillation phenomena.

This second test case deserves a specific focus: the ANN-based approach is capable to give a correct estimate of the LFO parameters also when the DMD method is not able to do so. This happens when NaN values (coming from the PMUs) are input to the DMD method. In correspondence of these special inputs there are no DMD outputs as shown in Figure 6a–c when purple markers are missing.

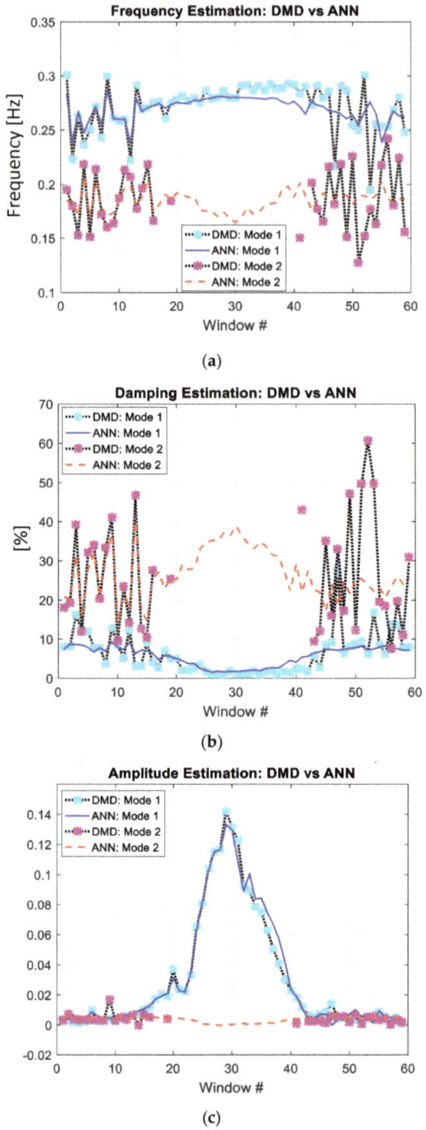

(a)

(b)

(c)

Figure 6. Comparison of the estimates produced by DMD versus the estimates produced by the ANN approach for the case of Dataset DS2: (**a**) frequency; (**b**) damping, (**c**) amplitude.

The last dataset to be explored in order to complete the analysis of validation scenario VS#1, is the one related to the 24 h recording of rated grid operations, namely DS3. In this particular test case the ANN has been trained on a data segment that is smaller than an half of the overall recording. This is due to the fact that, after several tests, the use of 50% or more of the data does not contribute to a better estimation. The training windows accounted for this scenario are only 50 out of 4381. The results are reported in Figure 7.

Figures from Figure 7a–c show the ANN and DMD results for all the 4381 windows. Although not easily distinguishable, they offer an overview of the overall trend of the ANN estimation and their comparison with the reference DMD. Figure 7d–f makes a focus on only 100 windows; also in this case the estimations provided by the ANN-based method closely follow those provided by the DMD method. The agreement between ANN and DMD results is very good for mode #1, the mode with less energy (blue line vs. light blue markers) and acceptable for the dominant mode #2 (dashed red line vs. purple markers). One aspect that should be underlined is that, even though there is a close tracking of the main trend of the estimates produced by the ANN with that of the DMD procedure, there are small difference between the two outputs. It is worthy to note that the estimations generated by the ANN-based approach are characterized by a smoother variation of the mode parameters, close to what happens in the real phenomenon.

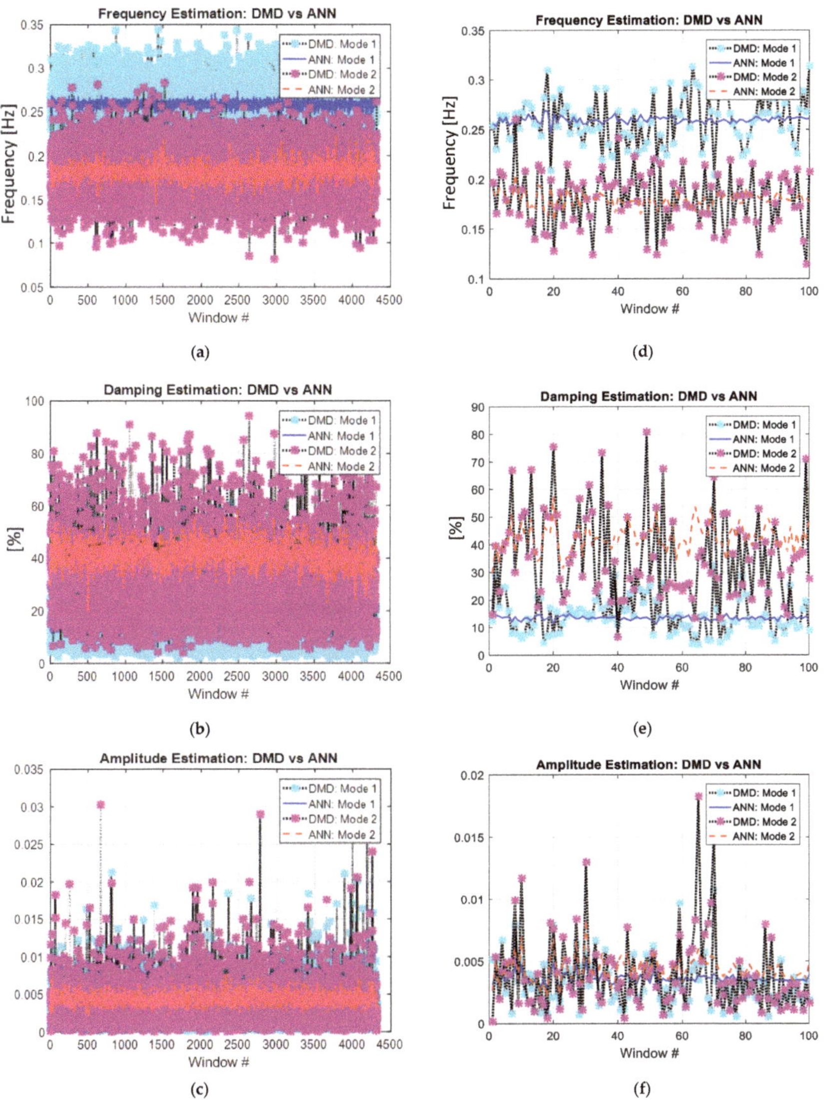

Figure 7. Comparison of the estimates produced by DMD versus the estimates produced by the ANN approach for the case of Dataset DS3: (**a**,**d**) frequency; (**b**,**e**) damping, (**c**,**f**) amplitude.

Because of the highly variable nature of the results in Figure 7, a further good figure of merit to assess the agreement between the ANN-based approach and the reference DMD is the mean value. Figure 8 shows the comparison of the ANN output for the dominant mode #2 (dashed red line) with the average value of the reference DMD output (purple thick line). The ANN estimates of the LFO mode #2 frequency and amplitude fit well the reference average (Figure 8a–c). The ANN estimates of the mode #2 damping factor (Figure 8b) is significantly off from the DMD average and, at this stage, no explanation is found.

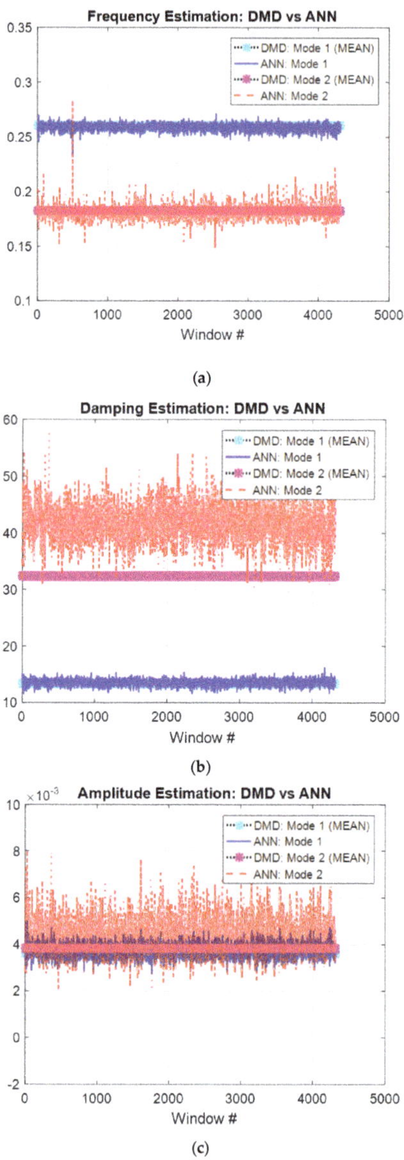

Figure 8. Comparison of the mean value of the estimates produced by DMD (purple thick line) versus the estimates produced by the ANN approach (dashed red line) for the case of DS3 and with a 20 s data frame: (**a**) frequency; (**b**) damping, (**c**) amplitude.

The aim of the validation scenario VS#2 is to assess the performances, in terms of LFO parameters, of different ANN architectures. Figure 9 shows the comparisons among the reference DMD results for the two modes (mode #1 light blue markers, mode #2 purple markers) and the corresponding ANN output for the three different ANN configurations described in Section 3.1 and Table 3.

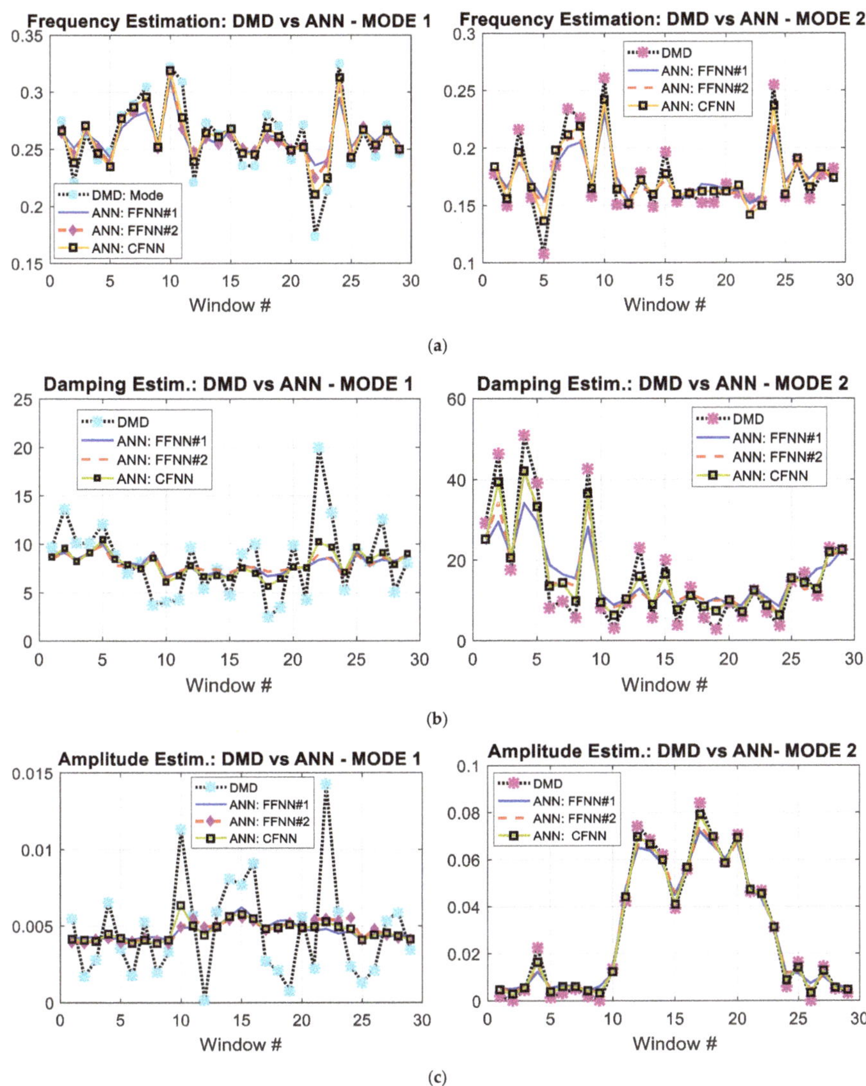

Figure 9. Comparison of the results obtained on the same dataset with different ANN architectures, according to the validation scenario VS#2: (**a**) frequency, (**b**) damping, (**c**) amplitude for mode #1 (left column) and mode #2 (right column).

The aim of the validation scenario VS#2 is to assess the performances in the estimation of the LFO parameters of different ANN architectures. Figure 9 shows the comparisons among the reference DMD results for the two modes (mode #1 light blue markers—left column, mode #2 purple markers—right column) and the corresponding ANN output for the three different ANN configurations described in Section 3.1 and Table 3. The visual inspection of the graph offers an immediate perception of the general agreement between the output results of the proposed ANN architectures and the DMD results with a better degree of accuracy for the mode #2. For a quantitative evaluation, the root-mean-square error (RMSE) between the reference DMD results and the results from each architecture and for each mode has been computed and reported in Table 4.

Table 4. Calculated root-mean-square error (RMSE) for the different ANN architectures of validation scenario VS#2.

Architecture	Estim. Mode	Calculated RMSE (Frequency)	Calculated RMSE (Damping)	Calculated RMSE (Amplitude)	Mean RMSE
FFNN#1	1	0.0202	0.0163	0.0030	0.1267
FFNN#1	2	0.0189	0.0324	0.0050	0.1781
FFNN#2	1	0.0178	0.0169	0.0030	0.1205
FFNN#2	2	0.0166	0.0246	0.0041	0.1430
CFNN	1	0.0129	0.0150	0.0028	0.0968
CFNN	2	0.0121	0.0174	0.0024	0.1024

4.3. Impact of the Data Window Lenght

The last validation scenario VS#3 is dedicated to the assessment of the effects, on the output results, of using different lengths of data windows. In this section the proposed ANN-based approach is implemented using time windows of 30 s and 60 s of DS1. The results obtained are reported in Figure 10. Basically, the use of a time frame of 30s does not heavily affect the ability of the ANN to recognize the three LFO parameters and, as expected, they match quite reasonably with those produced by the DMD method. However, two side effects should be considered in this study. First, the ANN performance outside the training set is degraded as the time window increases. Since the number of data is constant, longer time windows means less training set to be used in the training stage. Second, also the DMD results are negatively affected by this increase of time length. Although the demonstrated capability of both ANN and DMD to identify the correct dominant mode (mode #2) frequency at around 0.15 Hz (Figure 10a), the computed amplitude of mode #2 tends to be very close to that of mode #1 (see Figure 10c). This makes more difficult to identify the dominant mode. When each data window becomes longer than 60 s (Figure 10d–f) or more, the accuracy decreases even further.

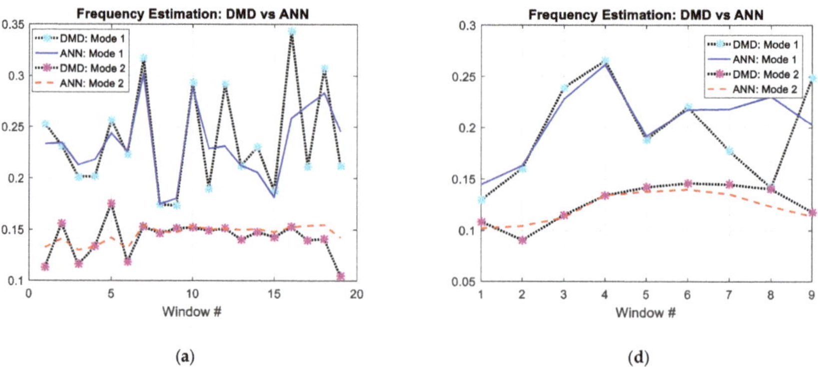

Figure 10. *Cont.*

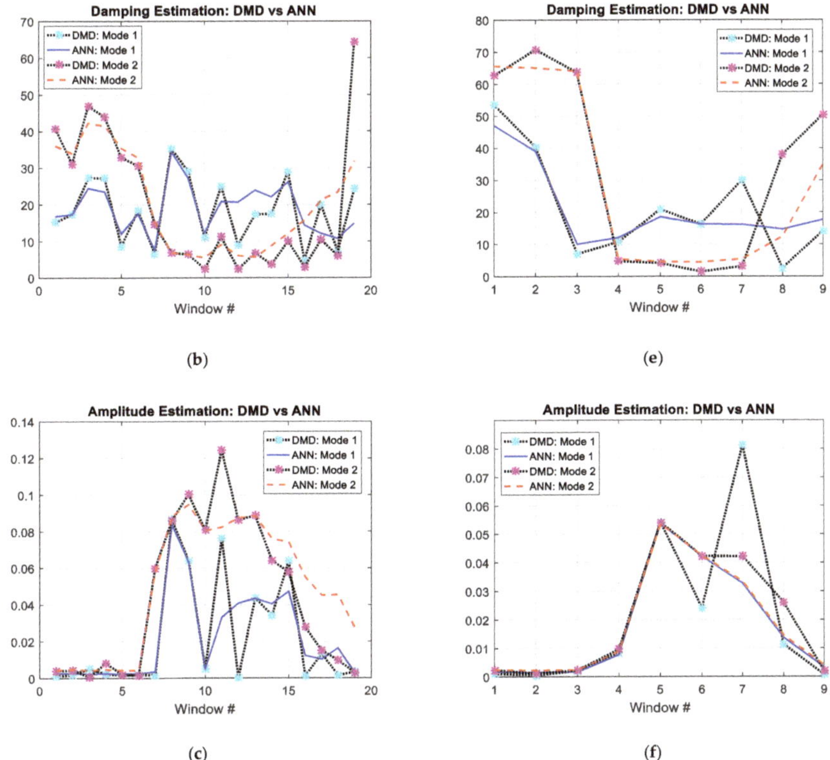

Figure 10. Estimates produced by DMD versus the estimates produced by the ANN approach for the case of dataset DS1 for a length of the data frame of 30 s: (**a**) frequency, (**b**) damping, (**c**) amplitude. Same comparisons evaluated for a length of the data frame of 60 s: (**d**) frequency, (**e**) damping, (**f**) amplitude.

5. Conclusions

This paper is the first step of a more extended project whose final aim is the development of an ANN architecture able to predict the system behavior (in a given time span) in terms of LFO modal parameters, and to classify the contingencies/disturbances based on an online training that has the memory of the passed training samples.

This contribution presented an ANN-based approach to estimate the mode parameters of LFO phenomena. The proposed technique is based on the development, offline training, and use of a suitably developed ANN architecture. The input and training data are real grid PMU frequency measurements provided by the Italian TSO. The development of the ANN training set is done through a preliminary pre-processing stage and by adopting a target vector selection policy, that is necessary in order to eliminate the detrimental effects induced by the occurrence of missing information in the PMU data stream in the form of NaN values.

The proposed technique has been validated using three main validation scenarios, in order to study the effectiveness of the method to recognize the LFO parameters, evaluate the best ANN architecture to be used, and assess the impact of data frame length.

From the obtained results it follows that the proposed method is capable to estimate, with a good degree of confidence, the three main LFO parameters for test case scenarios related to three real grid-recorded events. In this context, it seems that the best architecture to be used is the cascade feed-forward one, which offers the estimation with the lowest RMSE values.

In the perspective of the final target of the abovementioned project, the presented approach represents the proof of concept that the estimation and identification of LFO modal parameters from real PMUs measurement data streams can be reliably and efficiently performed by suitable ANN architectures, still trained offline, for each grid event under consideration. Overcoming this critical point is the actual object of the ongoing research efforts.

Author Contributions: Conceptualization and methodology, C.O., C.P., G.G., and A.O.; software and validation, C.O., C.P., G.G., F.d.P., and A.O.; resources, G.G.; project administration, C.P., R.S., and R.Z.; writing—original draft preparation, C.O., F.d.P., and A.O. All authors have read and agreed to the published version of the manuscript.

Funding: This research is funded by TERNA S.p.A., V.le Egidio Galbani, 70, 00156 Rome, Italy.

Conflicts of Interest: The authors declare no conflict of interest.

References

1. Ustinov, S.M.; Milanovic, J.V.; Maslennikov, V.A. Inherent dynamic properties of interconnected power systems. *Int. J. Electr. Power Energy Syst.* **2002**, *24*, 371–378. [CrossRef]
2. Ni, H.; Heydt, G.T. Power system stability agents using robust wide area control. *IEEE Trans. Power Syst.* **2002**, *17*, 1123–1131. [CrossRef]
3. Grigsby, L.L. *The Electric Power Engineering Handbook—Power System Stability and Control*, 2nd ed.; CRC Press, Taylor & Francis Group: New York, NY, USA, 2007.
4. El-Naggar, K. On-line measurement of low-frequency oscillations in power systems. *Measurement* **2009**, *42*, 716–721. [CrossRef]
5. Leonowicz, Z.; Lobos, T.; Rezmer, J. Advanced spectrum estimation methods for signal analysis in power electronics. *IEEE Trans. Ind. Electron.* **2003**, *50*, 514–519. [CrossRef]
6. Zeineldin, H.; Abdel-Galil, T.; El-Saadany, E.; Salama, M. Islanding detection of grid connected distributed generators using TLS-ESPRIT. *Electr. Power Syst. Res.* **2007**, *77*, 155–162. [CrossRef]
7. Izacard, G.; Mohan, S.; Fernandez-Granda, C. Data-driven Estimation of Sinusoid Frequencies. In Proceedings of the 33rd Conference on Neural Information Processing Systems (NeurIPS 2019), Vancouver, BC, Canada, 8–14 December 2019.
8. Viti, V.; Petrucci, C.; Barone, P. Prony methods in NMR spectroscopy. *Int. J. Imaging Syst. Technol.* **1997**, *8*, 565–571. [CrossRef]
9. Gudmundson, E.; Wirfält, P.; Jakobsson, A.; Jansson, M. An esprit-based parameter estimator for spectroscopic data. In Proceedings of the IEEE Statistical Signal Processing Workshop (SSP), Ann Arbor, MI, USA, 5–8 August 2012; pp. 77–80.
10. Quinn, B.G. Estimating parameters in noisy low frequency exponentially damped sinusoids and exponentials. In Proceedings of the 2016 IEEE International Conference on Acoustics, Speech and Signal Processing (ICASSP), Shanghai, China, 20–25 March 2016; pp. 4298–4302.
11. Adalbjörnsson, S.I.; Swärd, J.; Jakobsson, A. High resolution sparse estimation of exponentially decaying two-dimensional signals. In Proceedings of the 2014 22nd European Signal Processing Conference (EUSIPCO), Lisbon, Portugal, 1–5 September 2014; pp. 491–495.
12. Nouri-Sedeh, Z.; Mojiri, M.; Zekri, M. An adaptive linear neural network for identification of oscillatory damped signals. In Proceedings of the 21st Iranian Conference on Electrical Engineering (ICEE), Mashhad, Iran, 14–16 May 2013; pp. 1–5.
13. Quin, Q.; Gu, C. Prony based on-line identification and neural network control of drive systems. In Proceedings of the 2009 International Conference on Electrical Machines and Systems, Tokyo, Japan, 15-18 November 2009; pp. 350–357.
14. Philip, J.G.; Jain, T. Analysis of low frequency oscillations in power system using EMO ESPRIT. *Int. J. Electr. Power Energy Syst.* **2018**, *95*, 499–506. [CrossRef]
15. Poon, K.P.; Lee, K.C. Analysis of transient stability swings in large interconnected power systems by Fourier transformation. *IEEE Trans. Power Syst.* **1988**, *3*, 1573–1581. [CrossRef]

16. Hauer, J.F. Application of Prony analysis to the determination of modal content and equivalent models for measured power system response. *IEEE Trans. Power Syst.* **1991**, *6*, 1062–1068. [CrossRef]
17. Kumaresan, R.; Tufts, D.W. Estimating the parameters of exponentially damped sinusoids and pole-zero modeling in noise. *IEEE Trans. Acoust. Speech Signal Process.* **1982**, *30*, 833–840. [CrossRef]
18. Korba, P.; Larsson, M.; Rehtanz, C. Detection of oscillations in power systems using Kalman filtering techniques. In Proceedings of the 2003 IEEE Conference on Control Applications, 2003, (CCA 2003), Istanbul, Turkey, 25 June 2003.
19. Bronzini, M.; Bruno, S.; de Benedictis, M.; la Scala, M. Power system modal identification via wavelet analysis. In Proceedings of the 2007 IEEE Lausanne Power Tech, Lausanne, Switzerland, 1–5 July 2007; pp. 2041–2046.
20. Lauria, D.; Pisani, C. A two-step procedure for on-line detection of power oscillations. *Int. Rev. Electr. Eng.* **2012**, *7*, 4936–4947.
21. Messina, A.R.; Vittal, V. Nonlinear, non-stationary analysis of inter-area oscillations via Hilbert spectral analysis. *IEEE Trans. Power Syst.* **2006**, *21*, 1234–1241. [CrossRef]
22. Laila, D.S.; Messina, A.R.; Pal, B.C. A refined Hilbert–Huang transform with applications to interarea oscillation monitoring. *IEEE Trans. Power Syst.* **2009**, *24*, 610–620. [CrossRef]
23. Lauria, D.; Pisani, C. On Hilbert transform methods for low frequency oscillations detection. *IET Gener. Transm. Distrib.* **2014**, *8*, 1061–1074. [CrossRef]
24. Lauria, D.; Pisani, C. Improved non-linear least squares method for estimating the damping levels of electromechanical oscillations. *IET Gener. Transm. Distrib.* **2015**, *9*, 1–11. [CrossRef]
25. Giannuzzi, G.; Lauria, D.; Pisani, C.; Villacci, D. Real-time tracking of electromechanical oscillations in ENTSO-e Continental European Synchronous Area. *Int. J. Electr. Power Energy Syst.* **2015**, *64*, 1147–1158. [CrossRef]
26. Gupta, A.K.; Verma, K.; Niazi, K.R. Wide-area PMU-ANN based monitoring of low frequency oscillations in a wind integrated power system. In Proceedings of the 2018 8th IEEE India International Conference on Power Electronics (IICPE), Jaipur, India, 13–15 December 2018; pp. 1–6.
27. Gupta, A.K.; Verma, K. PMU-ANN based real time monitoring of power system electromechanical oscillations. In Proceedings of the 2016 IEEE 1st International Conference on Power Electronics, Intelligent Control and Energy Systems (ICPEICES), Delhi, India, 4–6 July 2016; pp. 1–6.
28. Zhao, Q.; Su, X.; Zhou, S. Research of power system stabilizer based on prony on-line identification and neural network control. In Proceedings of the 2008 International Conference on Electrical Machines and Systems, Wuhan, China, 17–20 October 2008; pp. 146–150.
29. Gupta, A.K.; Verma, K.; Niazi, K.R. Power system low frequency oscillations monitoring and generator coherency determination in real time. In Proceedings of the 2018 IEEE Innovative Smart Grid Technologies—Asia (ISGT Asia), Singapore, 22–25 May 2018; pp. 752–757.
30. Sulla, F.; Måsbäck, E.; Samuelsson, O. Linking damping of electromechanical oscillations to system operating conditions using neural networks. In Proceedings of the 2014 IEEE PES Conference on Innovative Smart Grid Technologies, Instanbul, Turckey, 12–15 October 2014; pp. 1–6.
31. Xiaomeng, L.; Venayagamoorthy, G.K. A neural network based wide area monitor for a power system. In Proceedings of the IEEE Power Engineering Society General Meeting, San Francisco, CA, USA, 16 June 2005; Volume 2, pp. 1455–1460.
32. Anaparthi, K.K.; Chaudhuri, B.; Thornhill, N.F.; Pal, B.C. Coherency Identification in Power Systems through Principal Component Analysis. *IEEE Trans. Power Syst.* **2005**, *20*, 1658–1660. [CrossRef]
33. Schmid, P. Dynamic mode decomposition of numerical and experimental data. *J. Fluid Mech.* **2010**, *656*, 5–28. [CrossRef]
34. Tu, J.H.; Rowley, C.W.; Luchtenburg, D.M.; Brunton, S.L.; Kutz, J.N. *On Dynamic Mode Decomposition: Theory and Applications*; American Institute of Mathematical Sciences: Springfield, MO, USA, 2014; Volume 12, pp. 391–421.
35. Kutz, J.N. *Data-Driven Modeling & Scientific Computation: Methods for Complex Systems & Big Data*; Oxford University Press: New York, NY, USA, 2013.

36. Chen, K.K.; Tu, J.H.; Rowley, C.W. Variants of dynamic mode decomposition: Boundary condition Koopman and Fourier analyses. *J. Nonlinear Sci.* **2012**, *22*, 887–915. [CrossRef]

Publisher's Note: MDPI stays neutral with regard to jurisdictional claims in published maps and institutional affiliations.

Article

Automatic P2P Energy Trading Model Based on Reinforcement Learning Using Long Short-Term Delayed Reward

Jin-Gyeom Kim and Bowon Lee *

Department of Electronic Engineering, Inha University, Incheon 22212, Korea; jg.kim@dsp.inha.ac.kr
* Correspondence: bowon.lee@inha.ac.kr; Tel.: +82-32-860-7423

Received: 3 September 2020; Accepted: 5 October 2020; Published: 14 October 2020

Abstract: Automatic peer-to-peer energy trading can be defined as a Markov decision process and designed using deep reinforcement learning. We consider prosumer as an entity that consumes and produces electric energy with an energy storage system, and define the prosumer's objective as maximizing the profit through participation in peer-to-peer energy trading, similar to that of the agents in stock trading. In this paper, we propose an automatic peer-to-peer energy trading model by adopting a deep Q-network-based automatic trading algorithm originally designed for stock trading. Unlike in stock trading, the assets held by a prosumer may change owing to factors such as the consumption and generation of energy by the prosumer in addition to the changes from trading activities. Therefore, we propose a new trading evaluation criterion that considers these factors by defining profit as the sum of the gains from four components: electricity bill, trading, electric energy stored in the energy storage system, and virtual loss. For the proposed automatic peer-to-peer energy trading algorithm, we adopt a long-term delayed reward method that evaluates the delayed reward that occurs once per month by generating the termination point of an episode at each month and propose a long short-term delayed reward method that compensates for the issue with the long-term delayed reward method having only a single evaluation per month. This long short-term delayed reward method enables effective learning of the monthly long-term trading patterns and the short-term trading patterns at the same time, leading to a better trading strategy. The experimental results showed that the long short-term delayed reward method-based energy trading model achieves higher profits every month both in the progressive and fixed rate systems throughout the year and that prosumer participating in the trading not only earns profits every month but also reduces loss from over-generation of electric energy in the case of South Korea. Further experiments with various progressive rate systems of Japan, Taiwan, and the United States as well as in different prosumer environments indicate the general applicability of the proposed method.

Keywords: automatic P2P energy trading; Markov decision process; deep reinforcement learning; deep Q-network; long short-term delayed reward

---

1. Introduction

In energy markets, the number of prosumers, i.e., the entities that generate and consume electric energy, has been increasing owing to the proliferation of distributed energy resources (DERs), such as photovoltaic (PV) systems, owned by traditional energy consumers. Accordingly, the proportion of microgrids in the power system has been expanding. In response, the incorporation of information and communication technology into existing power grids is becoming more important, and the core technologies of smart grid systems such as energy storage systems (ESSs), power conversion systems, mobility, and energy monitoring systems have advanced dramatically. In addition, studies on peer-to-peer (P2P) energy sharing or trading based on these core technologies between prosumers have been increasing [1–11], among which studies on P2P trading based on reinforcement learning (RL) [12] are actively being conducted [13–16]. Chen and Su [13] highlighted to the role of energy brokers in the localized event-driven market (LEM) because small-scale electricity consumers and prosumers typically take a long time to search for trading partners, and, therefore, pure P2P mode is not suitable. Nevertheless, these brokers aim to maximize profits in the LEM and determine the optimal action by using the Q-learning algorithm of RL [13]. In addition, it was suggested that the LEM participation strategy for energy trading can be modeled as a Markov decision process (MDP) and solved through a deep Q-network (DQN) [14]. Similarly, Chen and Bu [15] proposed a solution to the decision-making problem of microgrids in the LEM through a DQN-based P2P energy trading model of deep RL (DRL), and Liu et al. [16] applied a DQN for autonomous agents in the consumer-centric electricity market.

As such, recent studies on P2P energy trading define the strategy for participating in the trading as an MDP and apply RL or DRL to find the optimal trading participation strategy. This is because it is important to choose a reasonable and effective trading strategy in P2P energy trading. Therefore, the performance of RL, which is responsible for presenting the trading strategy, is very important for automatic P2P trading. However, most previous works [13–16] only applied RL or DRL to solve the MDP on energy trading, but did not consider the network modification of RL as they only considered the characteristics of energy trading to effectively solve it.

We aim to maximize the profits of the prosumer through automatic P2P energy trading, which is the same as that of stock trading algorithms. Therefore, we use the RL-based automatic trading algorithm used in stock trading to implement the automatic energy trading model. Our model provides optimal trading action based on independent prosumer ESS information, electricity generation, and consumption information for each designated trading time unit. Assuming that there exists a mechanism for the physical transaction of the P2P energy trading results, we present an implementation of an RL-based trading model for the automation of P2P energy trading and an effective network configuration by considering the unique factors of P2P energy trading.

In this paper, we propose a long short-term delayed reward (LSTDR) method that improves the existing delayed reward method of the RL network. LSTDR is a method that utilizes both short-term and long-term delayed rewards, enabling effective analysis of the long-and short-term patterns of trading environment information. To effectively analyze time-dependent information, we use a DQN based on a long short-term memory (LSTM) as the training model. The proposed method focuses on maximizing individual prosumer's profit based on noncooperative game theory [17] without considering the optimization of the overall benefits of all prosumers, so it is not directly related to Pareto optimality [18,19]. It does not consider the gain of consumers who do not generate electric energy either. Nevertheless, individual prosumers can benefit from adopting the proposed trading strategy at the same time reduce the overall energy generation of the grid which may potentially benefit consumers as well.

The remainder of this paper is organized as follows. Section 2 discusses the background information of the global energy market and P2P energy trading based on DRL and the existing works. Section 3 explains the difference between stock trading and energy trading, discusses the schemes for the modification of the automatic trading network by considering them, and proposes a new evaluation criterion for the trading strategy of LSTDR for RL. Section 4 presents the trading environment and experimental data for the performance evaluation of our proposed model. Section 5 discusses the experimental results. Finally, Section 6 concludes this paper.

2. Background and Related Works

2.1. Global Energy Market

Most countries fall into the category of energy producers which can produce energy from energy sources such as coal, oil and gas, or from renewable energy sources (RES). Energy produced in each country covers each domestic energy demand, and additionally needed or remaining energy after production is imported or exported to other countries, which activates the global energy markets [20]. In the scenario of the International Energy Agency (IEA)'s "World Energy Outlook 2019 (https://www.iea.org/reports/world-energy-outlook-2019)", the demand for primary energy in the global energy market will continue to increase every year, led by emerging economies such as China and India, while demand for primary energy will decrease in the developed countries, while the share of renewable energy will increase gradually for a low carbon emission to cope with the climate change. In addition, among the energy markets, the demand for electricity in the global electricity market will show the similar pattern, and the proportion of renewable energy in the supply is expected to increase significantly. The expansion of supply and demand for renewable energy in the electricity market may lead to the advancement of renewable energy generation technologies and the spread of supply [21], and increase the number of prosumers participating in the energy market for small-scale electricity generation through RES.

Each country provides benefits through various policies to promote electricity consumers to become prosumers. Most developed countries provide an environment where prosumers can generate profits by reducing electricity bills by increasing their self-consumption rate through self-generated electricity, or by selling it [22]. However, since the methods of electricity rate systems applied to each country are not all the same, even if prosumers in different countries have the same amount of consumption and generation, their profits can differ. For example, in two different countries with progressive rate system, if one country has a narrow range of progressive stages and a higher progressive rate compared to the other, the prosumers included in this country may gain less profits than those in the other country, even if they have less consumption and more electricity generation. This is an example of South Korea and the United States. In South Korea, the rate at the first progressive stage is lower than in the United States, but the progressive range is very narrow compared to the United States and the rate of increase in progression is much higher, resulting in higher electricity bills in South Korea compared to the same amount of electricity used in summer. Thus, the strategies of prosumers participating in trading in the electricity market should vary from country to country.

2.2. Peer-to-Peer Energy Trading

In P2P energy trading, the main agent is the prosumer, who produces and consumes energy and exchanges with other prosumers for surplus electricity that is overproduced after consumption [23]. Such P2P energy trading takes place in small DERs such as dwellings, factories, schools, and offices [6,7]. Unlike in the indirect trading method of conventional energy trading, where trading is performed through brokers offering wholesale or bundled services, in P2P energy trading, prosumers can trade directly with other prosumers (or consumers). Underscoring the strength of P2P energy trading, Tushar et al. [4] suggested that the development of this type of trading can lead to potential benefits for prosumers, such as earning profits, reducing electricity bills, and lowering their dependency on the grid. They also mentioned the importance of modeling the prosumer's decision-making process, noting that the system for energy trading requires reasonable modeling of each participant's decision-making process that can generate greater benefits for the entire energy network while considering human factors such as rationality, motivation, and environmental affinity for the trading. Therefore, it is important for P2P energy trading to set the direction and to model a strategy, and game theory [24] can be applied to this. Game theory can be divided into two main concepts: noncooperative game theory [17] and cooperative game theory [25]. In P2P energy trading, a noncooperative game sets a strategy with the goal of maximizing its own profits without the need to share and collaborate with other prosumers participating in the trading during the decision-making process. In contrast, in a cooperative game, for the benefit of all independent prosumers, they become the subject, share strategies and coordinate their own strategy choices. Therefore, even in the same energy trading environment, the game theory applied according to the purpose of the prosumer is different. In this study, we model the trading strategy on the basis of the noncooperative game theory that maximizes profits from the individual prosumer's point of view. Figure 1 compares the structures of non-P2P energy trading and the P2P energy trading.

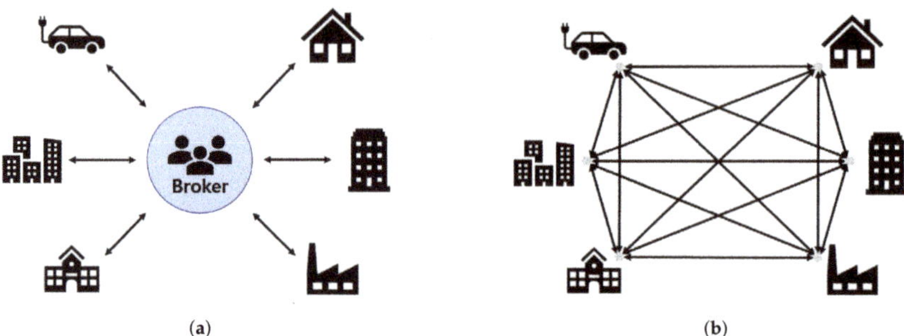

Figure 1. Comparison of the structures of (**a**) non-P2P energy trading and (**b**) P2P energy trading.

2.3. Markov Decision Process

The MDP is a discrete time probability control process that mathematically models and analyzes decision making, and it is designed according to the first-order Markov assumption that the current state is affected only by the previous state [26]. P2P energy trading is a decision-making problem in which a decision to participate in a trading can be defined as an MDP in an environment containing high-dimensional information. The MDP can be defined by the following five elements:

- State Space $\boldsymbol{S}$: This is the set of states s of the agent, which is the decision maker of a given environment E. In P2P energy trading, it is the state of the prosumer, which is the agent in a given trading environment, and includes the environmental information of the prosumer, such as energy generation, consumption, and energy reserves.
- Action Space $\boldsymbol{A}$: This is the set of all actions that the agent can select in a given state s. In P2P energy trading, the set of actions that a prosumer can select in a given state includes the actions for participating in the trading, such as buy, sell, and hold.
- Reward $\boldsymbol{R}$: This is the reward that the agent obtains from each action in a given state. Reward is typically a scalar, and depending on how the condition is set in each state, the reward obtained by the agent varies. There are two types of rewards: immediate reward, which is rewarded for the outcome of the next state, and delayed reward, which is rewarded for future results that are affected by the current behavior [27].
- State Transition Probability Matrix $\boldsymbol{P}$: State transition probability is the probability of transitioning from one state to another (or to the same state) and P is the matrix that defines the state transition probabilities in all states.
- Discount Factor γ: This is an element that plays a role in making the reward value of the future viewed from the present smaller according to the time distance from the present, considering that the future behavior is less affected by the present state as it returns with time. It has a value between 0 and 1.

Finally, a policy π to solve the MDP can be expressed through the above five elements and can be obtained through dynamic programming [28] or RL. Figure 2 shows the basic process of the MDP.

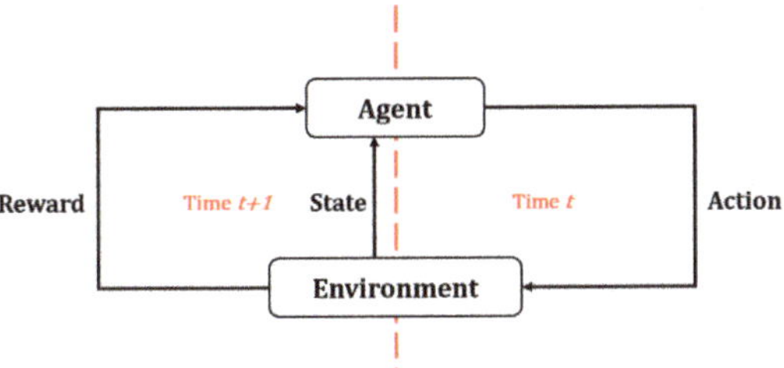

Figure 2. Markov decision process.

2.4. Deep Reinforcement Learning

DRL is a method that utilizes deep learning (DL) in RL algorithms [29]. RL algorithms optimize a policy using various approaches to find the optimal policy for the given goal. Representative algorithms of RL include SARSA [30], policy gradient [31], and Q-learning [32], and these algorithms update the main learning parameters using a function approximator to find the optimal policy [33]. DRL optimizes the policy by replacing this function approximator with DL. Accordingly, it is possible to effectively learn from a huge amount of data, and this has the advantage of improving the learning performance by applying various DL methods.

2.4.1. Deep Q-Networks

DQN [34] is a DRL algorithm that combines a deep neural network (DNN) with a Q-learning algorithm of RL. For Q-learning, a policy is recorded in the Q-table so that it can output the optimal action in each state of the agent [32]. However, this tabular recording of policy requires more memory as the amount of data increases or the dimension of the data increases. To solve this problem, function approximator is used to define the Q-function through parameters other than the table. The DQN uses the DNN as the function approximator [34] and applies the experience replay method to improve the data learning efficiency. To reduce the inefficiency of learning due to the correlation of adjacent learning data, the experience replay method stores the information about the agent's actions and the resulting state changes and rewards as a tuple-type transition in a buffer called replay memory and uses it for sampling during training. Therefore, it is possible to prevent a situation that falls into a local minimum by randomly selecting and using the transitions obtained in various environments during training. Algorithm 1 shows the overall algorithm structure of DQN proposed in [34].

Algorithm 1 Deep Q-learning with Experience Replay

```
Initialize replay memory D to capacity N
Initialize action-value function Q with random weights
for episode = 1, M do
    Initialise sequence s_1 = {x_1} and preprocessed sequence φ_1 = φ(s_1)
    for t = 1, T do
        With probability ε select a random action a_t
        otherwise select a_t = max_a Q*(φ(s_t), a; θ)
        Execute action a_t in the emulator and observe reward r_t and input x_{t+1}
        Set s_{t+1} = s_t, a_t, x_{t+1} and preprocess φ_{t+1} = φ(s_{t+1})
        Store transition (φ_t, a_t, r_t, φ_{t+1}) in D
        Sample a random minibatch of transitions (φ_j, a_j, r_j, φ_{j+1}) from D
        Set y_j = { r_j                                   for terminal φ_{j+1}
                  { r_j + γ max_{a_0} Q(φ_{j+1}, a_0; θ)   for non-terminal φ_{j+1}
        Perform a gradient descent step on (y_j − Q(φ_j, a_j; θ))^2
    end for
end for
```

2.4.2. Long Short-Term Memory

LSTM is a recurrent neural network (RNN) model in DL; it is effective for analyzing time-series data and is used for solving the gradient vanishing problem that occurs in vanilla RNN [35]. LSTM has a structure in which one memory cell c_t and three gates (i.e., input gate i_t, forget gate f_t, and output gate o_t, all at time t) that control the information flow are added to the vanilla RNN structure, so that long-term information can be effectively handled. Each gate operates in a different role [36]. The input gate determines how much the current information is reflected and stored in the memory cell, and the forget gate determines how much past information is forgotten and transferred to the memory cell. The output gate determines how much information is reflected and outputs the information currently stored in the memory cell. The operation of these gates is determined by an activation function σ (typically sigmoid or hyperbolic tangent) in each state. In this way, it is possible to effectively deal with time-series information because the information is updated by determining the importance and association of the information over time at each gate. The overall operating structure of LSTM can be expressed by the following equations [35]:

$$i_t = \sigma_g(w_i x_t + U_i h_{t-1} + b_i), \tag{1a}$$

$$f_t = \sigma_g(w_f x_t + U_f h_{t-1} + b_f), \tag{1b}$$

$$o_t = \sigma_g(w_o x_t + U_o h_{t-1} + b_o), \tag{1c}$$

$$c_t = f_t \circ c_{t-1} + i_t \circ \sigma_c(w_c x_t + U_c h_{t-1} + b_c), \tag{1d}$$

$$h_t = o_t \circ \sigma_h(c_t). \tag{1e}$$

For the input sequence $\mathbf{x} = \{x_1, x_2, x_3, ..., x_T\}$ in Equation (1), x_t represents the input at time t; U_i, U_f, U_o, U_c, w_i, w_f, w_o and w_c are the weight matrices; and b_i, b_f, b_o, and b_c are the bias vectors, all of which are the parameters that are updated during training. Finally, c_t and hidden layer output h_t, which is the information transmitted to the next state, are calculated through the Hadamard product ($\circ$), which is the element-wise product of the output of each gate and information c_{t-1} and h_{t-1} transmitted from the previous state. Figure 3 shows the architecture of LSTM.

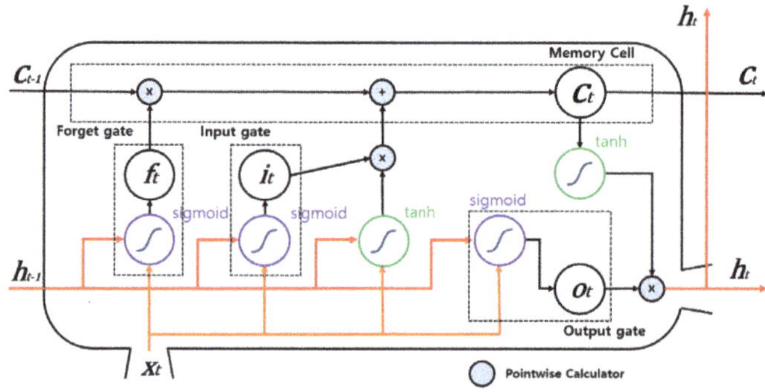

Figure 3. Architecture of LSTM.

3. Proposed Approach

Stock trading is the buying and selling of stocks. It is a nonphysical type of trading as there is no exchange of physical products. In stock trading, the agents participating in the trading aim to maximize their gains through trading at the optimum time using the market price of stocks that fluctuate in real time. Accordingly, whether or not the agents will participate in the trading mostly depends on the market price of the stock. Figure 4 shows the algorithm of automatic stock trading based on DRL.

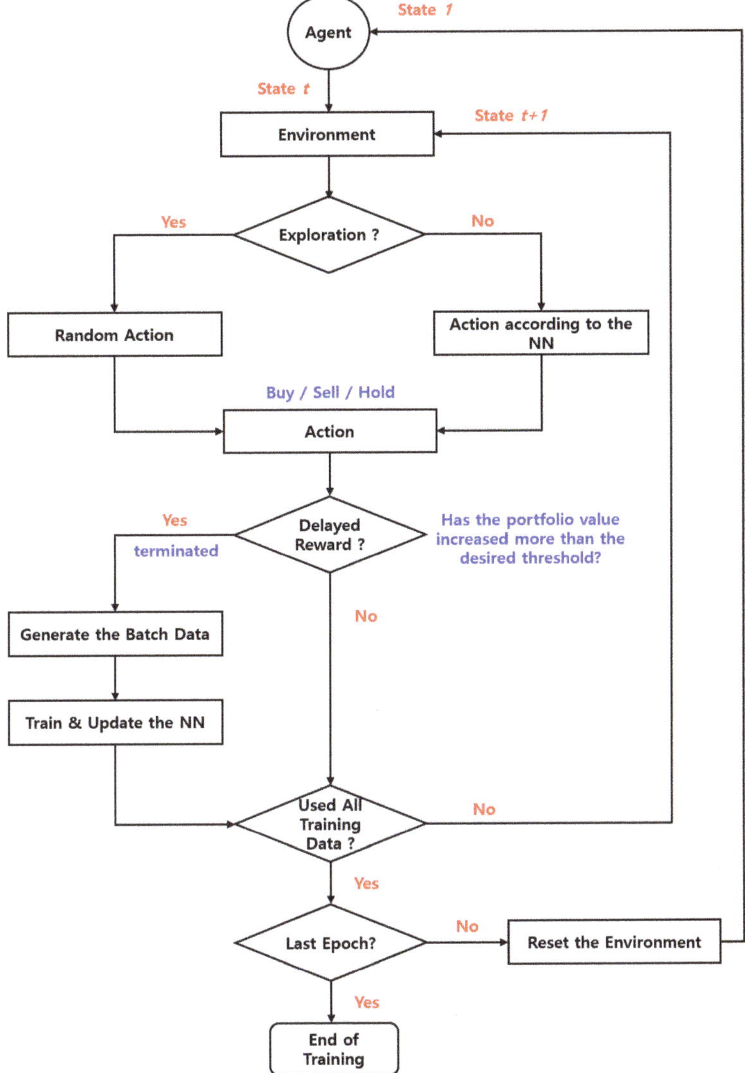

Figure 4. Stock trading algorithm based on deep reinforcement learning.

Unlike stock trading, energy trading has additional factors that affect the trading conditions. First, the electricity subject to energy trading can be generated and consumed by the prosumers; therefore, its reserves can change in real time even when they are not traded, unlike stocks whose reserves change only by trading. Therefore, it is necessary to redefine the evaluation criterion of energy trading to make it different from that of stock trading, which uses the portfolio of the sum of only the currency values of the assets held. Second, unlike stocks, which are virtually traded, electricity is physically traded, and, therefore, there is a trading time until a trading is made and terminated. Third, when electricity is charged or discharged, losses occur depending on the ESS efficiency, and, therefore, the actual trading result will be different from the initial trading volume. Considering these, we propose strategies for designing a model suitable for automatic P2P energy trading. The automatic energy trading model adopts the existing automatic stock trading algorithm and modifies it to match the specifics of energy trading.

3.1. Definition of the P2P Energy Trading Evaluation Criterion

The electricity reserves in the ESS resulting from the energy generation and consumption of prosumer constantly change even if there is no trading, and the energy trading determines the trading action according to the reserves in the ESS changed by the previous trading, consumption, or generation. Therefore, if the currency value of the assets held in stock trading is used as an evaluation criterion of the trading, it is impossible to accurately evaluate the trading results owing to the changes in energy generation and consumption. Therefore, we define an evaluation criterion as the sum of the gains from participating in P2P energy trading compared with not participating in it.

The total profit from participating in the P2P energy trading proposed in this paper is defined as the sum of four gains. The first gain is the change in the electricity bill. When energy trading is completed, since the result of the trading changes the amount of electricity held in the ESS, the amount of electricity available to the prosumer in the ESS is changed, and the amount of electricity supply used is also changed accordingly. For example, if the amount of electricity held in the ESS is less than the amount consumed, the electricity bill can be reduced by purchasing electricity through P2P energy trading rather than through the supply electricity. Conversely, while the amount of electricity held in the ESS is greater than the amount consumed, selling through P2P energy trading can result in profit, although it may involve the use of supply electricity, which may increase the electricity bill. Therefore, if only the gain from the trading is considered without the electricity bill, there may be a situation in which additional electricity bills are paid more than the gain from the trading, resulting in loss. The gain from the change in electricity bill can be expressed as follows:

$$G_{bill}(S_p(t)) = B_o(S_o(t)) - B_p(S_p(t)), \tag{2}$$

where $t = 1, 2, 3, \ldots, T$; $B_o(S_o(t))$ is the electricity bill paid by the prosumer who does not participate in P2P energy trading in state $S_o(t)$ and $B_p(S_p(t))$ is the electricity bill paid by the prosumer who participates in P2P energy trading in state $S_p(t)$. In both situations, the difference in electricity bills is $G_{bill}(S_p(t))$, which is the gain from the change in the electricity bill for participating in energy trading, where t is the number of states that have elapsed from the start of the electricity bill calculation to the hour-by-hour period, and T is the total number of states from the time the final electricity bill is calculated. We assume that the time before the trading is established, the electricity is transferred, and the ESS is completely charged/dischargid is within 1 h, thereby setting the trading participation decision interval to 1 h. Therefore, t increases in units of 1 h. The second gain is the trading gain from P2P energy trading. When a prosumer participates in a P2P electricity trading, the prosumer takes one of three actions: buying, selling, and nonparticipation, and the prosumer's assets change as a result of the trading. The amount of change in these assets is equal to the gain achieved only through trading, and it can be defined as

$$0 \le Q_b(S_p(t)) \le 1/\eta \cdot E_{max}, \tag{3a}$$

$$0 \le Q_s(S_p(t)) \le \eta \cdot E_{max}, \tag{3b}$$

$$M_b(S_p(t)) = (1+\xi) \cdot (P(S_p(t)) \cdot Q_b(S_p(t))), \tag{3c}$$

$$M_s(S_p(t)) = (1-\xi) \cdot (P(S_p(t)) \cdot Q_s(S_p(t))), \tag{3d}$$

$$M_{trade}(S_p(t)) = \sum_{k=1}^{t} M_s(S_p(k)) - \sum_{k=1}^{t} M_b(S_p(k)), \tag{3e}$$

$$G_{trade}(S_p(t)) = M_{trade}(S_p(t)) - M_{trade}(S_o(t)), \tag{3f}$$

$$G_{trade}(S_p(t)) = M_{trade}(S_p(t)), \tag{3g}$$

where E_{max} represents the maximum storage capacity of the ESS, and η represents the efficiency of the ESS. $Q_b(S_p(t))$ and $Q_s(S_p(t))$ represent the trading purchase and sale volume, respectively, and $P(S_p(t))$ represents the trading price. $M_b(S_p(t))$ and $M_s(S_p(t))$ represent the cost spent on purchases and the profits from sales, respectively. At this time, ξ represents the trading fee. $M_{trade}(S_o(t))$ and $M_{trade}(S_p(t))$, which are the total amount of asset changes through trading, are calculated as the difference between the total amount of revenue and the expenditure up to t. When the prosumer does not participate in P2P trading, the asset changes through trading $M_{trade}(S_o(t))$ are zero, and, therefore, the gain from the P2P energy trading $G_{trade}(S_p(t))$ is equal to $M_{trade}(S_p(t))$. Since the trading is based on the ESS, the amount of electricity that can be traded is limited. Therefore, the trading volume should consider the amount of electricity held in the ESS or the remaining storage capacity of the ESS, and it cannot exceed the maximum ESS capacity. In addition, trading fees may also be considered in P2P energy trading, which should be further considered in the settlement of trading costs. The third gain is for virtual losses from over-generation. If electricity is generated while the electricity in the ESS is fully charged, the generated electricity cannot be stored in the ESS, resulting in losses. However, this can be prevented through the sales from P2P energy trading before these losses occur. Therefore, it is possible to perform an efficient trading action considering the electricity generation by the prosumer and the losses from over-generation depending on whether or not P2P energy trading is involved. The gain on virtual losses from over-generation can be expressed as follows:

$$V_{gain}(S_p(t)) = \eta \cdot (L_o(S_o(t)) - L_p(S_p(t))) \cdot P(S_p(t)), \tag{4a}$$

$$G_{virtual}(S_p(t)) = \sum_{k=1}^{t} V_{gain}(S_p(k)), \tag{4b}$$

where $L_o(S_o(t))$ and $L_p(S_p(t))$ are the amount of electricity loss from over-generation, and $V_{gain}(S_p(t))$ is the instantaneous gain on the currency value of the virtual loss in state $S_p(t)$. $G_{virtual}(S_p(t))$ is the cumulative gain from reducing over-generation. Setting up a trading strategy in such a way as to reduce losses from over-generation not only can reduce the losses of prosumers but also can have the effect of reducing the total amount of supply electricity on the power system. The fourth gain is the change in the currency value of the amount of electricity held in the ESS. The electricity held in the ESS is the result of consumption, generation, and trading, and it includes the result of the changes due to the trading actions. The gain from the change in the currency value of the amount of electricity in the ESS can be expressed as follows:

$$C_g(S_{o,p}(t)) = \eta \cdot Generation, \tag{5a}$$

$$D_c(S_{o,p}(t)) = -1/\eta \cdot Consumption, \tag{5b}$$

$$C_b(S_p(t)) = \eta \cdot Q_b(S_p(t)), \tag{5c}$$

$$D_s(S_p(t)) = -1/\eta \cdot Q_s(S_p(t)), \tag{5d}$$

$$E(S_o(t)) = E(S_o(t-1)) + C_g(S_o(t)) + D_c(S_o(t)), \tag{5e}$$

$$E(S_p(t)) = E(S_p(t-1)) + C_g(S_p(t)) + D_c(S_p(t)) + C_b(S_p(t-1)) + D_s(S_p(t-1)), \tag{5f}$$

$$G_{ess}(S_p(t)) = \eta \cdot (E(S_p(t)) - E(S_o(t))) \cdot P(S_p(t)), \tag{5g}$$

where $C_g(S_{o,p}(t))$ and $D_c(S_{o,p}(t))$ respectively represent the amount of ESS charged and discharged owing to the generation and consumption of the prosumer from state $S_{o,p}(t)$ to state $S_{o,p}(t-1)$. Moreover, $C_b(S_p(t))$ and $D_s(S_p(t))$ respectively represent the amount of electricity charged and discharged by the prosumer through P2P energy trading. $E(S_o(t))$ and $E(S_p(t))$ represent the amount of electricity in the ESS. In addition, because the effect (charge or discharge) of the trading result in state

$S_p(t-1)$ is not immediately apparent but is shown in the next state $S_p(t)$, the amount of electricity in the ESS $E(S_p(t))$ after the prosumer's P2P energy trading utilizes the trading volume in the previous state $S_p(t-1)$ rather than the current trading volume in state $S_p(t)$. $G_{ess}(S_p(t))$ represents the gain from the currency value of the electricity held in the ESS. Finally, the profit $G(S_p(t))$ for participating in P2P energy trading, which has been redefined as the trading evaluation criterion, is defined as follows:

$$G(S_p(t)) = G_{bill}(S_p(t)) + G_{trade}(S_p(t)) + G_{ess}(S_p(t)) + G_{virtual}(S_p(t)). \quad (6)$$

3.2. Long Short-Term Delayed Reward

MDPs that have a continuous environment such as automatic trading do not have exact termination points, unlike MDPs that have an episode's termination point, such as mazes, Atari games, and CartPoles [37]. Therefore, in the case of the existing RL-based stock trading [38], the RL structure is designed to generate the termination points of learning by providing a delayed reward [27] when a certain amount of portfolio gain or loss is achieved during the trading. However, we considered various external factors affecting energy trading when defining the evaluation criterion for energy trading in Section 3.1 and utilized the gain from electricity bills. The electricity bill is set according to the amount of electricity used each month. Accordingly, we aimed to determine the monthly gains, and, for this purpose, we set the time that the delayed reward is the output of automatic energy trading at the end of the period when the electricity price is set so that the termination point of the episode is generated every month. Similarly, most of the papers on energy trading set the trading strategy by designating a certain size (period) for an episode [13,14,39,40], and the policy is updated by using the immediate rewards that occur in every state in episode and the delayed reward generated at the termination point of the episode. The delayed reward value only determines the end point of the episode to proceed with the policy update, but is not directly used for policy update. The policy update reflects the impact on the future by applying a discount factor to each of the immediate rewards from the current state to the state in which the delayed reward occurred.

Such a one-month delayed reward assignment, however, can make learning difficult for short-term patterns that occur within a month. To compensate for this, we design an additional delayed reward to include the case when an increase or decrease in the profit that we defined occurs above a certain threshold as in the stock trading method. At this time, the delayed reward is not provided whenever an increase or decrease occurs above a certain threshold, such as in stock trading, but is added to the final delayed reward by utilizing the number of occurrences of the increase or decrease within a month. Finally, the delayed reward information is utilized when deciding on the action in each state to ensure that the outcome of a month's trading affects the learning direction in the training. For this, the obtained delayed reward is added to the Q-function updated through the neural network. By applying the delayed reward method of stock trading to energy trading, we enable the DNN in RL at the beginning of the learning to focus on very short-term patterns (because of the generation of batch data based on the delayed reward that occurs every short period of time) and then to find that the monthly pattern is important while finding the overall training direction only after a great deal of training has progressed. However, to do it, we can set the monthly term as a unit of training (which results in a delayed reward every month) so that we can train from the beginning of the training in a way that fits our goals. In addition, the ratio of the profit and the number of trades was utilized to add to the final delayed reward. By doing so, when delayed reward occurs owing to a profit above a threshold, we can direct the trading model to effectively maximize the profit while giving a higher score continuously to a situation where more gain is obtained instead of the same delayed reward score. Therefore, we propose an LSTDR method for delayed reward by considering long-term patterns of 1 month and short-term patterns in the long-term. Figure 5 shows the structure of the proposed automatic P2P energy trading scheme. Algorithm 2 shows the overall algorithm structure for the energy trading with LSTDR method.

Algorithm 2 Energy trading with the LSTDR method

Initialize replay memory D to capacity N
Initialize action-value function Q with random weights
for epoch = 1, M **do**
  Initialise sequence $s_1 = \{x_1\}$ and preprocessed sequence $\phi_1 = \phi(s_1)$
  Initialize G_{total} (total gain) to zero and G_{base} (base gain) to non-zero
  Initialize C_s (count value for the occurrence of non-zero R_s (short-term delayed reward)) to zero
  Initialize S_s (cumulative value of R_s) to zero
  Initialize P_m (cumulative value of the unit trading price) to zero
  Initialize N_{trade} (number of times prosumer participated in the trading) to zero
  Initialize N_s (count value for the number of states in a month) to zero
  Set g_s (threshold value of the short-term profit) to user's desired value (we set it to 0.2)
  Set g_m (threshold value of the monthly profit) to user's desired value (we set it to 0.25)
  for t = time of the first training data, time of the last training data **do**
    $N_s += 1$
    With probability ϵ select a random action a_t
    otherwise select $a_t = \max_a Q^*(\phi(s_t), a; \theta)$
    Execute action a_t in the emulator and observe immediate reward r_t and environment x_{t+1}
    Set $s_{t+1} = s_t, a_t, x_{t+1}$ and preprocess $\phi_{t+1} = \phi(s_{t+1})$
    Store transition $(\phi_t, a_t, r_t, \phi_{t+1})$ in D
    Update G_{total} with G_{bill}, G_{trade}, G_{ess} and $G_{virtual}$
    Update B_o (prosumer's current electricity bill when not participating in the P2P trading)
    Update P_m and N_{trade}
    if G_{base} is equal to or greater than zero **then**
      if G_{total} is equal to or greater than $(1 + g_s) \cdot G_{base}$ **then** R_s is 1
      else if G_{total} is greater than $(1 - g_s) \cdot G_{base}$ and less than $(1 + g_s) \cdot G_{base}$ **then** R_s is 0
      elseR_s is -1
      end if
    else
      if G_{total} is equal to or greater than $(1 - g_s) \cdot G_{base}$ **then** R_s is 1
      else if G_{total} is greater than $(1 + g_s) \cdot G_{base}$ and less than $(1 - g_s) \cdot G_{base}$ **then** R_s is 0
      elseR_s is -1
      end if
    end if
    $S_s += R_s$
    if R_s is not zero **then** $C_s += 1$ and $G_{base} = G_{total}$
    end if
    if t is the end of the month **then**
      $R_a = ((G_{total} - G_{virtual}) / N_{trade}) \cdot (N_s / P_m)$
      if $(G_{total} - G_{virtual})$ is equal to or greater than $B_o \cdot g_m$ **then**
        R_l (long-term delayed reward) is $R_a + 1$
      else if $(G_{total} - G_{virtual})$ is greater than zero and less than $B_o \cdot g_m$ **then** R_l is R_a
      elseR_l is $R_a - 1$
      end if
      R_{total} is $\alpha \cdot R_l + (1 - \alpha) \cdot (S_s / C_s)$ (α is a weight factor between 0 and 1)
      Initialize $G_{total}, G_{base}, C_s, S_s, P_m, N_s, N_{trade}$
    elseR_{total} is zero
    end if
    if R_{total} is non-zero **then**
      Sample a random minibatch of transitions $(\phi_j, a_j, r_j, \phi_{j+1})$ from D
      Set $y_j = \begin{cases} r_j + R_{total} & \text{for terminal } \phi_{j+1} \\ r_j + \gamma \max_{a_0} Q(\phi_{j+1}, a_0; \theta) + R_{total} & \text{for non-terminal } \phi_{j+1} \end{cases}$
      Perform a gradient descent step on $(y_j - Q(\phi_j, a_j; \theta))^2$
    end if
  end for
end for

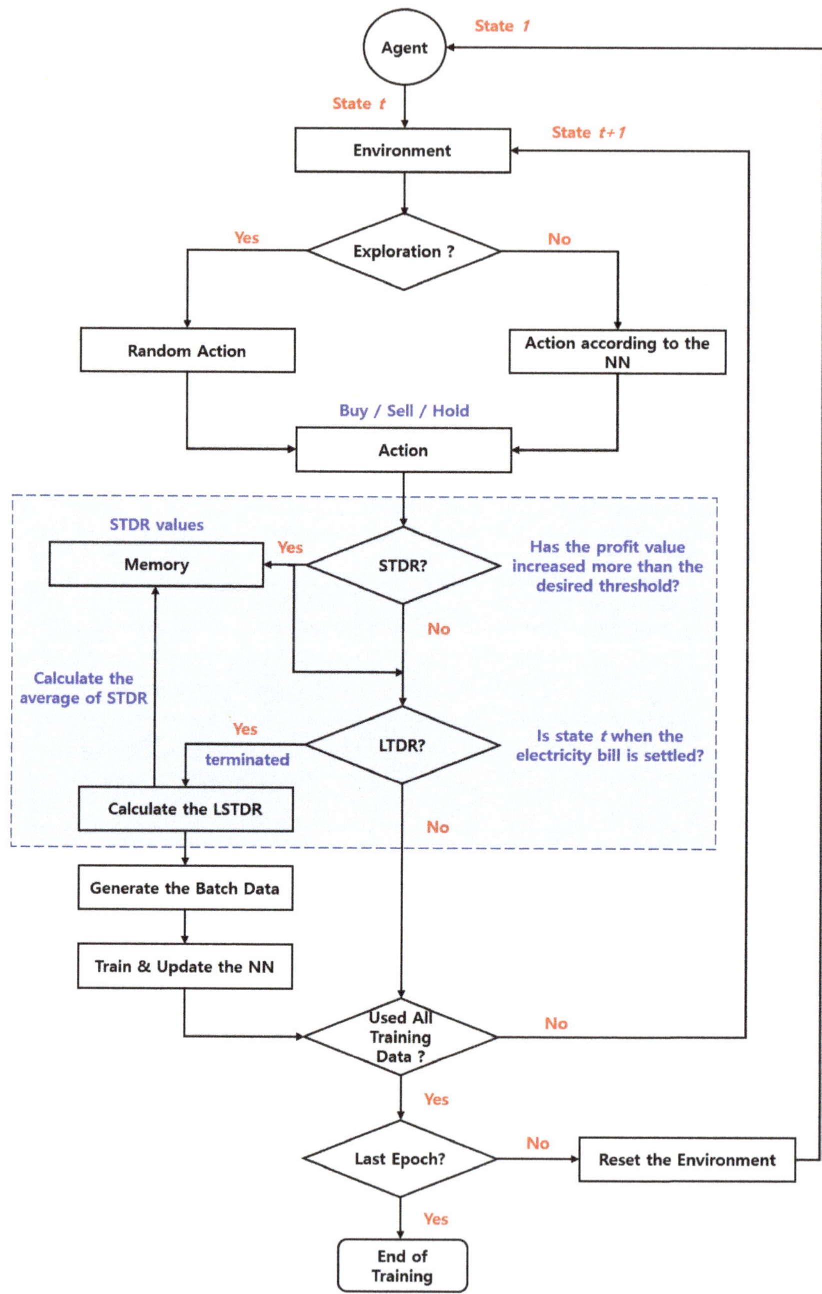

Figure 5. Energy trading algorithm based on LSTDR.

4. Experiments

In our work, we defined the trading environment for the evaluation of the proposed P2P automatic energy trading model and generated the experimental data accordingly. The experiment was conducted under the assumption that P2P energy trading exists in South Korea, and the public data in South Korea were utilized for the data generation. In the experiment, we verified the validity of the proposed LSTDR method and confirmed the profit that the prosumer would gain by participating in the trading through the proposed P2P automatic energy trading model. In the first experiment, we compared the results of three delayed reward methods: the short-term delayed reward (STDR) method [27], which is a delayed reward method used in stock trading; long-term delayed reward (LTDR) method [13,14,39,40], which generates a termination point every month to provide delayed reward at that time; and the LSTDR method, which utilizes both of these methods and is the one proposed in this paper. The second experiment compared the prosumers who did and did not participate in the trading to confirm the benefits of participating in P2P energy trading. In the last set of experiments, we confirmed whether the proposed P2P energy trading model is applicable to the various electricity rate system in other countries as well as the changes in the energy consumption and generation of the prosumers.

4.1. *Definition of the Trading Environment*

For the creation of the datasets and the conduct of the experiments, we first assumed that P2P energy trading exists in South Korea and prosumer was defined as a three- to four-person household whose electricity bill is set at the end of each month in the progressive rate system. Table 1 shows the information on the progressive rate system applied to the household.

Table 1. Information on the progressive rate system applied to the household.

Season	Consumption (kWh)	Basic Rate (USD)	Progressive Rate (USD/kWh)
Summer (July, August)	0–300	0.78	0.08
	301–450	1.37	0.16
	451–	6.23	0.24
Others	0–200	0.78	0.18
	201–400	1.37	0.24
	401–	6.23	0.28

In addition, it is assumed that the general household, which is a prosumer, has an ESS and can obtain information such as electricity generation and consumption in real time through a smart meter and the amount of electricity held in the ESS. As an external requirement, it is assumed that the distribution lines with other prosumers are connected in advance, so that there is no need to perform a follow-up work after the trading is completed, and that the full charging or discharging of electricity in the ESS is assumed to occur within 1 h after the trading is completed. The environmental factors and setup information based on these assumptions are listed in Table 2.

Table 2. Setup for the trading environment.

Environmental Factor	Setup Information
Country	South Korea
Prosumer	Three- to four-person household
Electricity bill	Progressive rate system
Generation method	PV system
Generator capacity	3 kW
Average daily consumption	Less than 10 kWh
Smart meter	Installed
ESS	Installed
ESS storage capacity	8 kWh

4.2. Definition of the Dataset

We generated a dataset according to the assumptions made in Section 4.1, and the generated dataset contains three types of information. The first information is time information. Time information is represented as month, day, and hour in three channels, all in integer form. Table 3 shows the definition of time information in the dataset.

Table 3. Definition of time information in the dataset.

Time Information	Definition	Range
Month information	Month information of the date, expressed as an integer starting from January to December.	1–12
Day information	Day of the week information of the date, expressed as an integer starting from Monday to Sunday.	1–7
Hour information	Hour information of the date, expressed as an integer in hours from 00:00 (midnight) to 23:00 hours (11:00 pm).	0–23

As described above, by providing the time information on a daily basis, we enable the neural network in RL to effectively learn the pattern information depending on the time of day in a trading environment. The second information is weather information because electricity generation and consumption are sensitive to weather conditions. Therefore, by using weather information, we could effectively predict the generation and consumption of prosumer, and to make trading decisions by considering this information. We used 21 types of weather information provided by the Korea Meteorological Administration (KMA) (https://data.kma.go.kr/data/grnd/selectAsosRltmList.do?pgmNo=36) on an hourly basis. Therefore, the weather information in the dataset consists of 21 parameters as listed in Table 4.

Table 4. Definition of weather information in the dataset.

Factor	Unit	Factor	Unit
Temperature	°C	Wind direction	°
Wind speed	m/s	Precipitation	mm
Humidity	%	Vapor pressure	hPa
Dew point	°C	Local atmospheric pressure	hPa
Sea-level pressure	hPa	Sunshine duration	h
Solar radiation	MJ/m^2	Snowfall	cm
Total Cloud	10 quantiles	Low-middle level cloud	10 quantiles
Height of lowest cloud	100 m	Visibility	10 m
Ground temperature	°C	(5 cm) Underground temperature	°C
(10 cm) Underground temperature	°C	(20 cm) Underground temperature	°C
(30 cm) Underground temperature	°C		

The final information is prosumer information, which consists of two dimension of prosumer electricity generation and consumption. We previously defined a prosumer as a general three- to four-person household and set it up to generate electricity only through a PV system. Based on this, the KMA's sunshine duration information was utilized to generate the virtual information for the electricity generation by the prosumer. In addition, demand forecast data for domestic pricing plans and average monthly electricity usage information for three- to four-person households were utilized to generate information on the virtual electricity consumption of the prosumer. Figure 6 shows the data for the generated virtual electricity consumption and generation.

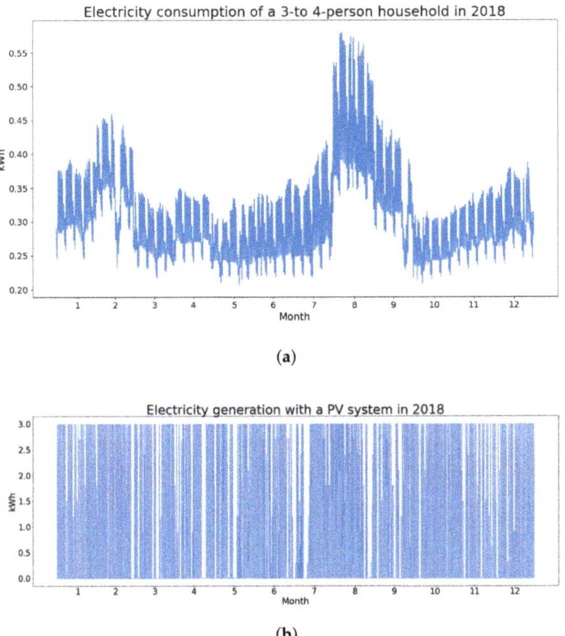

Figure 6. Generated data for electricity consumption and generation of a three- to four-person household: (**a**) consumption and (**b**) generation.

In addition, we used the electricity bill rate and the demand forecast data for the domestic pricing generation plan to generate the prosumer's desired trading price information in proportion to the electricity demand, and the generated data are shown in Figure 7.

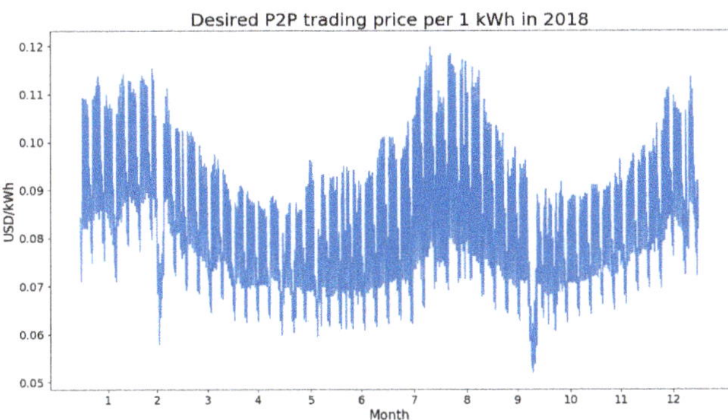

Figure 7. Generated data for the desired trading price of the prosumer.

As a result, the generated dataset consists of a total of 27 dimension, and we generated data for a total of 3 years from 2016 to 2018. Among them, the data for 2016 and 2017 were used for the training and those for 2018 were used for the testing.

4.3. Hyperparameters for Learning

In DRL, hyperparameters are the factors that affect the operation of various algorithms in the model; thus, they greatly affect the model performance. The hyperparameters in DQN we used as a trading model can be divided into hyperparameters for RL and those for DL, and our hyperparameter settings are shown in Table 5.

Table 5. Hyperparameter settings.

Algorithm	Hyperparameter	Value	Explanation
DL	Learning rate	0.0001	The value that affects the optimization speed of the DL model during training. It determines the degree of update of the parameter in the process of reaching the optimal point for the learning goal.
	Hidden layer size	256	The number of nodes in the hidden layer in the DL model. The input data of the hidden layer is expressed as a new feature value for each node.
	Optimizer	Adam	Parameter optimization model. It allows parameters to be updated in a direction that matches the goal during training.
	Epoch	200	A unit of learning in which all data from the training dataset are used.
RL	Replay memory size	8760	The number of transitions that can be stored in DQN's replay memory. The number of states is equivalent to one year (24 × 365).
	Epsilon	0.5	The probability of random exploration in the decision of action. It enables the exploration of various environments so that they can learn strongly about environmental changes. The epsilon value decreases as the epoch increases in the training of DL.
	Discount factor	0.99	The value that reflects the degree of future impact on the reward.

4.4. Experiment Environment

We used TensorFlow and Keras as the DRL framework for the experiment and a workstation with a high-performance GPU. The details are as follows:

- CPU: Intel(R) Xeon(R) Silver 4110 CPU @ 2.10GHz (32 cores)
- GPU: Tesla V100
- OS: Ubuntu 16.04.5 LTS
- TensorFlow Version : 1.14.0
- Keras Version : 2.3.1

5. Results and Discussion

5.1. Validation of the Proposed Delayed Reward Method

In Section 3.2, we presented the difficulty of applying the STDR method used in stock trading as a delayed reward method for the energy trading model and the approaches to compensate for it. To verify the effectiveness of the proposed method, we compared the results by applying the following methods to each trading model: the STDR method; the LTDR method, which is a delayed reward method that considers the electricity bills; and the LSTDR method, which complements the LTDR method, as the latter cannot learn short-term patterns well. This experiment utilized the contents and dataset defined in Section 4. Figure 8 shows the patterns of the monthly profit change for each delayed reward method.

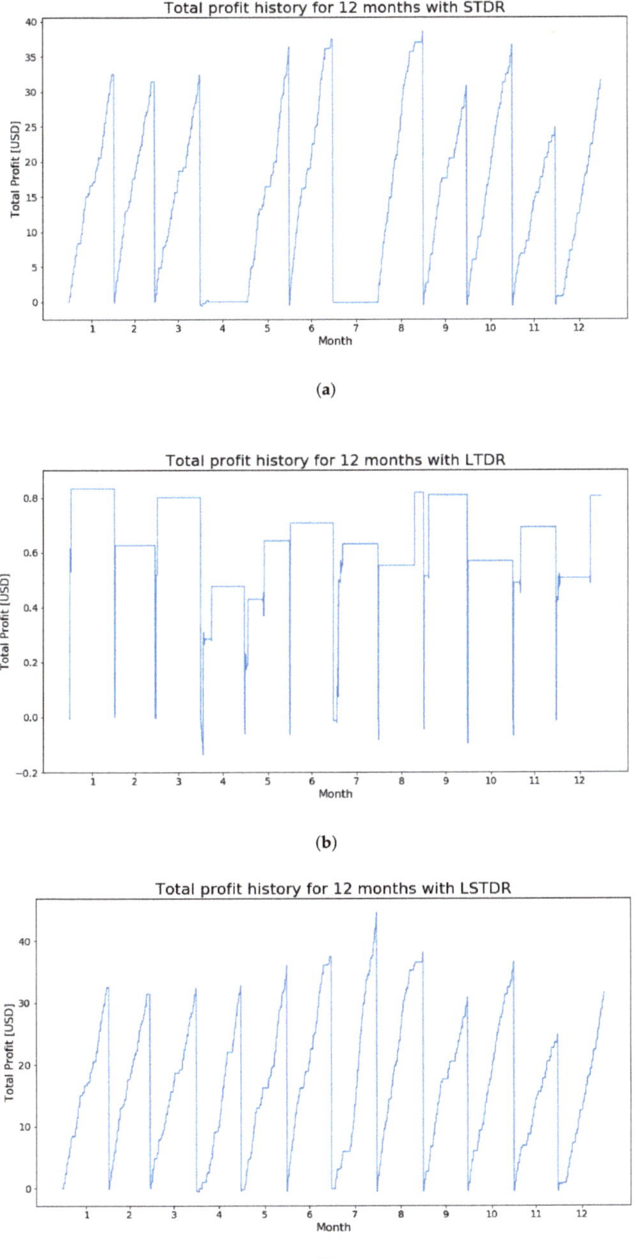

(a)

(b)

(c)

Figure 8. Comparison of the patterns of the monthly profit change according to the delayed reward method: (**a**) STDR, (**b**) LTDR, and (**c**) LSTDR.

Figure 8a shows the change in profit for monthly trading when STDR is used as a delayed reward method. The change in monthly profit generally shows a pattern of steady increase. This is because there is no designated termination point of a episode, and when a profit change over a certain threshold occurs, the episode can be terminated to learn various short-term patterns. However, because there is no designated termination point of a episode, if the model is not well trained for short-term patterns through sufficient exploration, it may not generate profits every month. The results showed that most, but not all, of the months have high profits. On the contrary, in Figure 8b, which is the result of using the LTDR method, it can be seen that profits are generated in all months, but are much less than those of the STDR method. This can generate a profit for each month because the LTDR method generates the termination point for an episode every month; however, it is difficult to learn short-term patterns because delayed reward occurs once a month. Therefore, it can be seen that the model has not been trained in the direction of steadily increasing profit. Figure 8c shows the result of solving the problems in the previous two delayed reward methods and of achieving a higher profit every month. Therefore, it is concluded that a more effective energy trading model can be generated through LSTDR, which scores the number of STDRs occurring within a month and reflects them in the results of the LTDR method.

5.2. Comparison of Gains from Participating in P2P Energy Trading

We defined profit in Section 3.1 as a criterion for evaluating the trading results by including various gains. In this section, the gains of participating in P2P energy trading are identified and the resulting profit is finally identified. The experiment was conducted using the content and dataset defined in Section 4 based on the proposed LSTDR energy trading model. Figure 9 shows a comparison of the electricity bills of a consumer who does not generate and consumes only, a non-trading prosumer who does not participate in the trading, and a trading prosumer who participates in the trading.

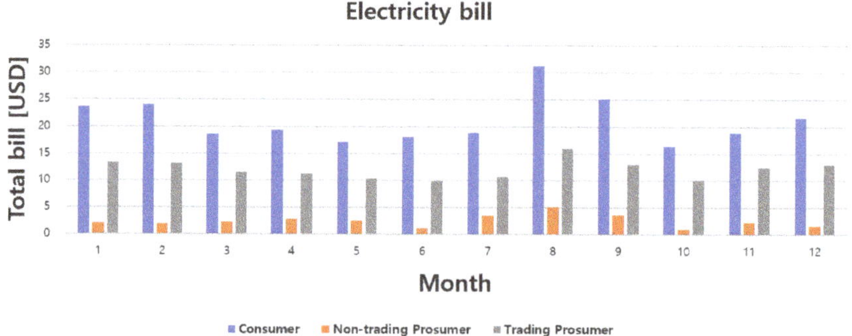

Figure 9. Monthly electricity bills.

Figure 9 shows the result of the prosumer's participation in P2P energy trading, where it pays additional electricity bills. This is because if the prosumer does not participate in the trading, most consumption can be covered by the electricity stored in the ESS after the electricity generation; however, if it participates in the trading and sells it, the number of situations in which the consumption cannot be covered through electricity in the ESS increases and, therefore, the amount of supply electricity used increases. The reason for this trading strategy is that the trading price is higher than the supply electricity price. In the generation of the experimental dataset, we did not generate the prosumer's trading price separately for the purchase and sales, but, instead, we generated it as one trading price in proportion to the electricity demand. The trading price generated in this way is higher than the first-stage rate of the progressive rate and is lower than the second-stage rate. Therefore, the prosumer tends to maximize gains by selling electricity in the ESS and using cheap supply electricity when the first-stage rate is applied. In response, when the prosumer participates in the trading, it shows the result of using the supply electricity in a way that does not go beyond the first stage of the progressive rate in every month.

We assumed that the prosumer sets the average daily consumption below 10 kWh and uses a PV system with a capacity of 3 kW. Therefore, daily electricity generation is larger than consumption. However, the ESS's capacity is set at 8 kWh, resulting in lost electricity from over-generation. For this situation, we mentioned that P2P energy trading can reduce the amount of electricity lost from over-generation, and the experimental results to confirm this are shown in Figure 10.

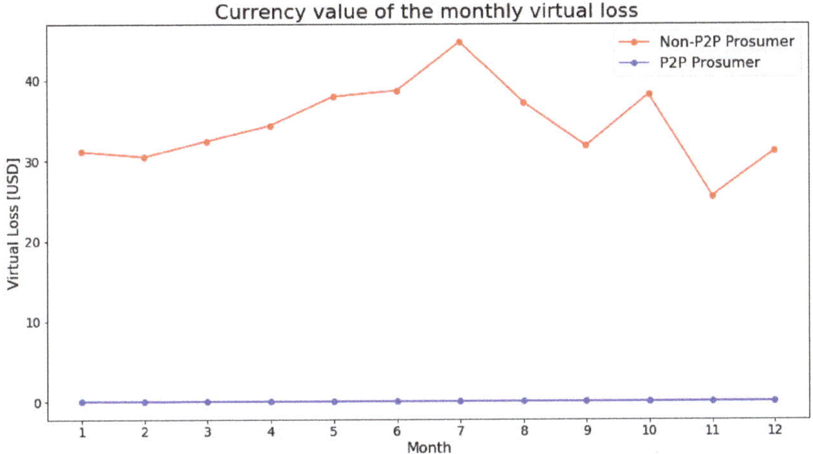

Figure 10. Currency value of the monthly virtual loss from over-generation.

Figure 10 shows the currency value of the monthly virtual loss arising from over-generation. If the prosumer does not trade in the over-generation condition, it loses a large amount of virtual losses per month. However, by participating in the trading, the prosumer can sell the electricity that is over-generated, and, therefore, electricity that is lost can be minimized. In addition, although this study has set the goal of generating a trading strategy for its own gains, the sale of electricity through energy trading may also result in further reduction of the use of electricity in the power grid. In this experiment, it was shown that it is possible to avoid the situation of over-generation through trading, thus showing no virtual loss every month.

We identified four gains that we defined to obtain a profit for participating in energy trading. Figure 11 shows the monthly gains of the prosumer by participating in energy trading.

As shown in Figure 11, prosumer tends to be more profitable when participating in P2P energy trading, although they must pay a higher monthly electricity bill; despite the latter, there are many gains from trading and a reduced virtual loss. In fact, the result of the gain on virtual loss is implied in the trading gain. This is because it has gained from the trading in as much as there was no loss. Therefore, when considering the profit to be earned at the end of each month, we do not need to additionally consider the gain on virtual loss. Figure 12 shows the monthly profit that the prosumer ultimately earns by participating in P2P energy trading.

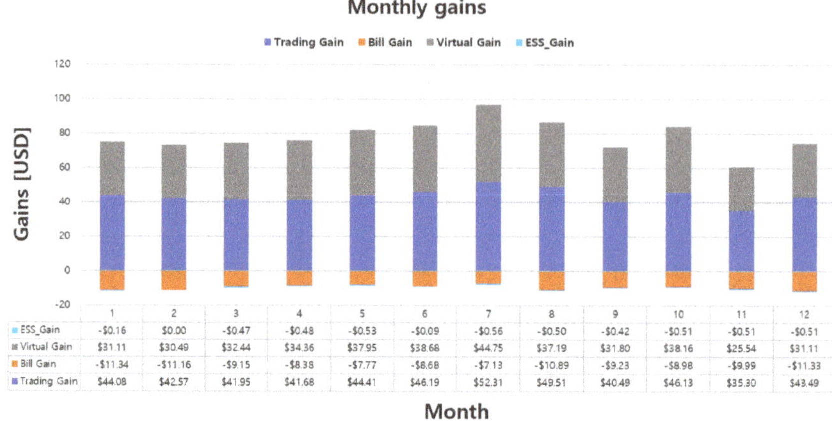

Figure 11. Monthly gains of the prosumer by participating in energy trading.

Figure 12. Monthly profit of the prosumer by participating in energy trading.

In Figure 12, total profit shows the profit including the gain on virtual loss and real profit shows the profit actually gained by prosumer as the sum of other gains except the gain on virtual loss. Prosumer has shown positive results for both profits by participating in P2P energy trading, and most profits have been obtained by selling lost electricity that cannot be stored after generation because the ESS capacity is fully charged.

5.3. Comparison of Various Rate Systems

We applied the proposed model to various rate systems based on the prosumer's trading environment defined in Section 4.1. First, South Korea does not adopt a fixed rate system, but in order to compare it with the progressive system applied in the previous experiment, the situation of the fixed rate system was defined and the results were confirmed. The rate of the fixed rate system was set to a value higher than the first-stage rate of the progressive system and lower than the second-stage rate, and Figure 13 shows the monthly total profits of the progressive and fixed rate system in the same trading environment.

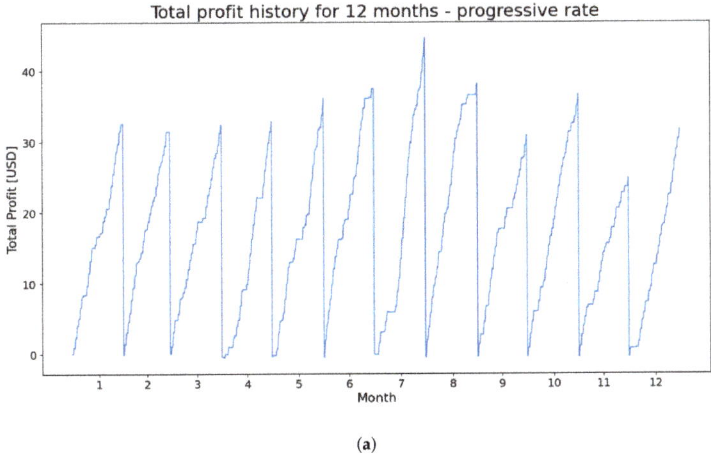

(**a**)

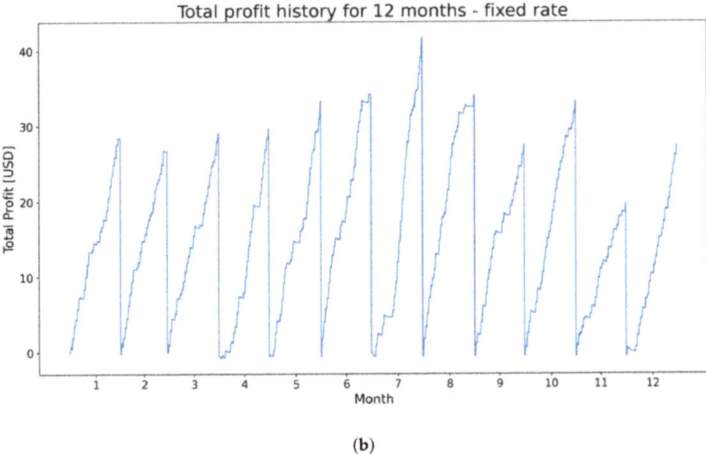

(**b**)

Figure 13. Monthly total profits comparison of (**a**) progressive rate system and (**b**) fixed rate system.

As shown in Figure 13, the patterns of the total monthly profits for the progressive and fixed rate systems are very similar, with the former being higher than the latter. This is because when the amount of generation is greater than the amount of consumption, the progressive rate does not change, and the progressive system is applied like a fixed rate system. However, it can be seen that the first-stage rate of the progressive system is lower than that of the fixed rate system, so that more profits are obtained.

Secondly, by applying the compositions of the progressive systems of other countries to the rate system, the monthly total profits were compared. We used the progressive rates of the United States, Taiwan, and Japan provided by Korea Electric Power Corporation (KEPCO) (http://cyber.kepco.co.kr/ckepco/front/jsp/CY/H/C/CYHCHP00302.jsp) to the experiment, which is shown in Table 6. Figure 14 shows the comparison of monthly total gains for each country and Figure 15 shows the pattern of changes in total monthly gains for each country.

Table 6. Information on the progressive rate system applied to the household in each country.

Country	Season	Consumption (kWh)	Progressive Rate (USD/kWh)
USA	All	0–1000	0.0915
		1001–	0.1002
Taiwan	Summer (June–September)	0–120	0.072
		121–330	0.10
		331–500	0.15
		501–700	0.19
		701–1000	0.21
		1001–	0.23
	Others	0–120	0.072
		121–330	0.092
		331–500	0.12
		501–700	0.15
		701–1000	0.17
		1001–	0.18
Japan	All	0–120	0.18
		121–300	0.24
		301–	0.28

Monthly total profits for each country

KOREA JAPAN TAIWAN USA

Profit [USD]

	1	2	3	4	5	6	7	8	9	10	11	12
KOREA	$32.58	$31.42	$32.34	$32.82	$36.11	$37.41	$44.62	$38.12	$30.84	$36.64	$24.81	$31.64
JAPAN	$48.80	$47.04	$50.93	$52.32	$59.11	$60.94	$74.01	$56.72	$47.19	$59.43	$36.68	$47.48
TAIWAN	$19.95	$19.22	$20.68	$21.22	$23.89	$24.64	$29.85	$22.44	$18.87	$24.03	$15.00	$19.39
USA	$22.06	$21.19	$22.64	$23.29	$26.15	$26.86	$33.17	$26.65	$21.41	$26.14	$16.32	$21.31

Month

Figure 14. Monthly profit of the prosumer by participating in energy trading.

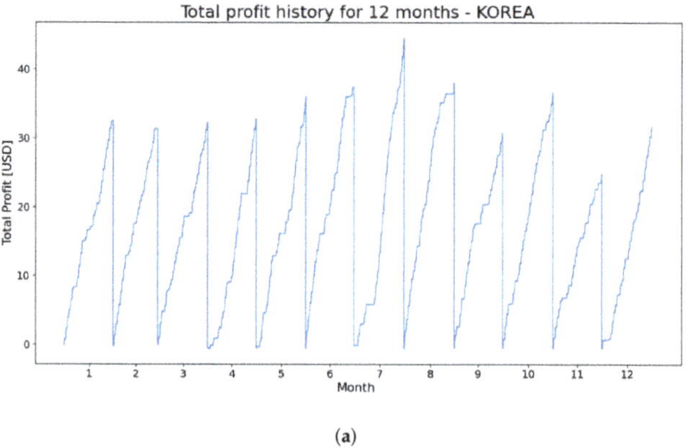

(a)

Figure 15. *Cont.*

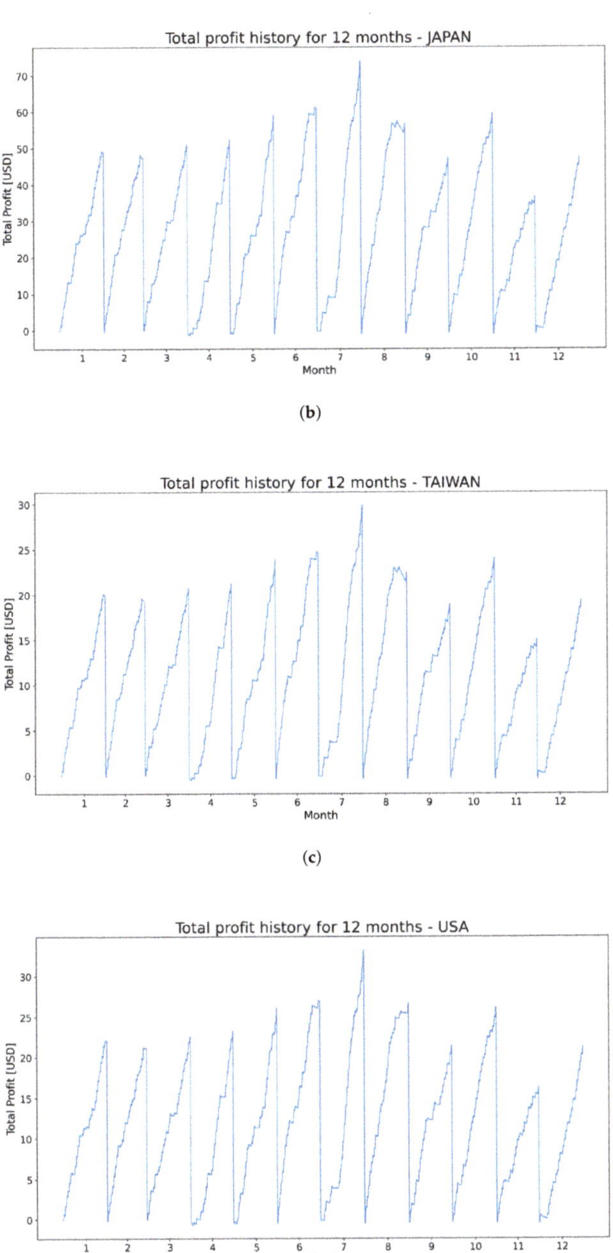

Figure 15. Monthly total profits comparison of (**a**) South Korea, (**b**) Japan, (**c**) Taiwan, and (**d**) USA.

As shown in Figure 14, proposed model was able to gain monthly profits even when applied to rate systems in various countries. Also, as shown in Figure 15, almost similar trading strategies are being created for prosumers with the same trading environment, because the progressive rate of all countries is fixed at the first-stage due to the amount of generation more than consumption. In this regard, in order to effectively confirm the trading strategy that changes according to the composition of the progressive rate system, we doubled the consumption of prosumers and changed the PV power generation capacity from 3 kWh to 750 Wh, so that progressive rate changes can occur well. Figure 16 shows the patterns of monthly total profits for the changed energy consumption and generation of the prosumer.

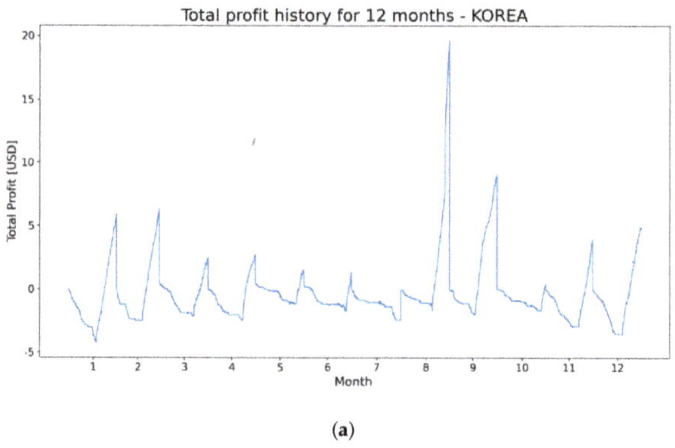

(a)

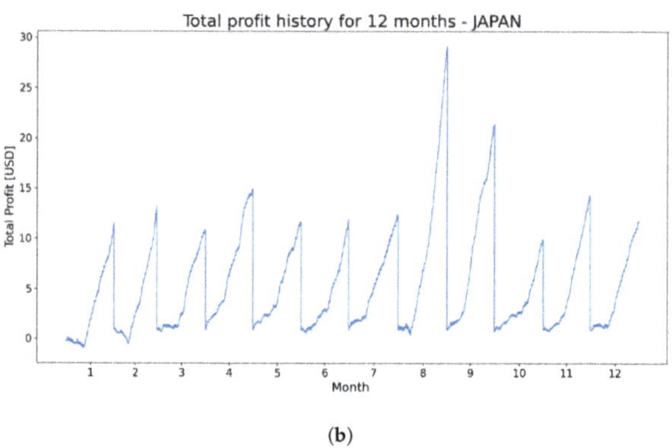

(b)

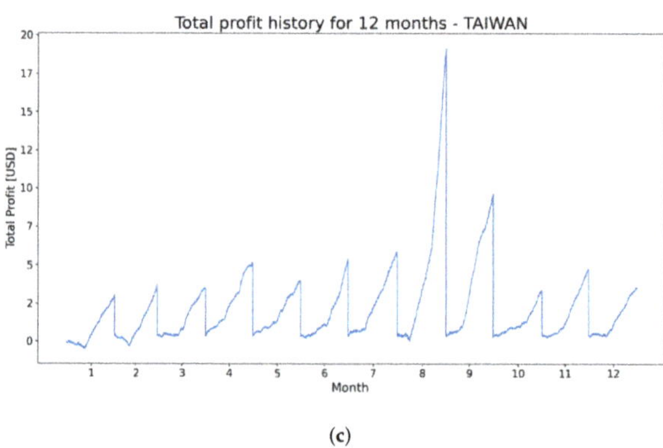

(c)

Figure 16. *Cont.*

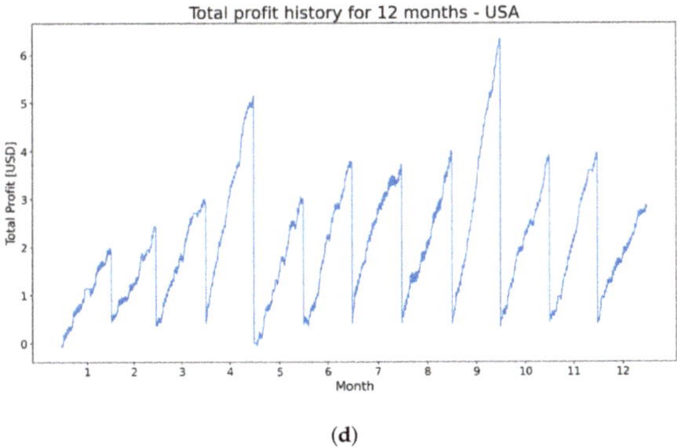

(d)

Figure 16. Monthly total profits comparison of (**a**) South Korea, (**b**) Japan, (**c**) Taiwan, and (**d**) USA.

As shown in Figure 16, it can be seen that different trading strategies are created according to the change of the trading environment to maximize profits. In this experiment, higher progressive rate is applied in countries other than the United States because of the higher consumption. As a result, even if most prosumers take loss in advance, they purchase electricity through tradings, and later benefit from staying in the lower bracket of the progressive rate of the billing system. This indicates that the proposed model is capable of creating different trading strategies for each country owing to the different progressive rate compositions. These results indicate that the proposed model can be applied well to various trading environments.

6. Conclusions

In this paper, we proposed an LSTM-based DQN model with the LSTDR method as an effective automatic P2P energy trading model; we also proposed a new evaluation criterion that can effectively learn the long-term and short-term patterns of the trading environment. We set the goal of a noncooperative game theory-based trading strategy that maximizes the prosumer's profit through participation in P2P trading. The profit is defined as the sum of four gains, and each gain is obtained by comparing the case where the prosumer does not participate with the case where the prosumer participates in the trading.

A comparative experiment was conducted with the STDR method, which is a delayed reward method used in stock trading, and the LTDR method, which can learn a specific long-term patterns of information by designating the termination point of the episode. By using the proposed LSTDR method, we were able to solve the problem of the STDR method, which does not obtain the profit every month, and the issue with the LTDR method, which can obtain a profit every month but not a large amount of it.

We set up a virtual energy trading environment by designating a three- to four-person household in South Korea that generates electricity through a PV system as a prosumer, and we conducted experiments using the LSTDR method-based energy trading model. In the experiment, we confirmed each of the gains we defined and finally confirmed the profits that prosumer would earn by participating in P2P energy trading. The proposed trading strategy tended to generate trading gains through continuous sales, without deviating from the progressive rate of the electricity bill that is cheaper than the trading price, resulting in losses due to the payment of additional charges for the electricity bill gain. Nevertheless, it was able to achieve the highest trading gain. In addition, by trading, the prosumer could reduce the amount of electricity lost from over-generation. The same trend can also be found with the fixed rate system. Finally, the prosumer was able to earn a profit every month, showing that it can benefit from participating in P2P energy trading.

Further experiments with different progressive rate systems in Japan, Taiwan and the United States as well as the changes of the energy consumption and generation of the prosumers indicate the general applicability of the proposed method. However, prosumers may belong to a variety of trading environments other than the rate systems, and may participate in trading for different purposes. In the future, we plan to build a trading environment that is closer to a real-world by considering the situation of various prosumers and enabling trading between prosumers through trading matching.

Author Contributions: B.L. and J.-G.K. conceived this paper.; J.-G.K conducted an experiment and analyzed the experimental results; B.L. and J.-G.K. wrote and edited this paper. All authors have read and agreed to the published version of the manuscript.

Funding: This research was funded by Korea Electric Power Corporation (Grant number: R18XA01) and the Institute of Information & communications Technology Planning & Evaluation (IITP) grant funded by the Korea government (MSIT) (2020-0-01389, Artificial Intelligence Convergence Research Center (Inha University)).

Conflicts of Interest: The authors declare no conflict of interest. The funders had no role in the design of the study; in the collection, analyses, or interpretation of data; in the writing of the manuscript, or in the decision to publish the results.

References

1. Etukudor, C.; Couraud, B.; Robu, V.; Früh, W.G.; Flynn, D.; Okereke, C. Automated negotiation for peer-to-peer electricity trading in local energy markets. *Energies* **2020**, *13*, 920. [CrossRef]
2. Alvaro-Hermana, R.; Fraile-Ardanuy, J.; Zufiria, P.J.; Knapen, L.; Janssens, D. Peer to peer energy trading with electric vehicles. *IEEE Intell. Transp. Syst. Mag.* **2016**, *8*, 33–44. [CrossRef]
3. Morstyn, T.; Farrell, N.; Darby, S.J.; McCulloch, M.D. Using peer-to-peer energy-trading platforms to incentivize prosumers to form federated power plants. *Nat. Energy* **2018**, *3*, 94–101. [CrossRef]
4. Tushar, W.; Yuen, C.; Mohsenian-Rad, H.; Saha, T.; Poor, H.V.; Wood, K.L. Transforming energy networks via peer-to-peer energy trading: The potential of game-theoretic approaches. *IEEE Signal Process. Mag.* **2018**, *35*, 90–111. [CrossRef]
5. Liu, Y.; Wu, L.; Li, J. Peer-to-peer (P2P) electricity trading in distribution systems of the future. *Electr. J.* **2019**, *32*, 2–6. [CrossRef]
6. Zhang, C.; Wu, J.; Zhou, Y.; Cheng, M.; Long, C. Peer-to-Peer energy trading in a Microgrid. *Appl. Energy* **2018**, *220*, 1–12. [CrossRef]
7. Zhang, C.; Wu, J.; Long, C.; Cheng, M. Review of existing peer-to-peer energy trading projects. *Energy Procedia* **2017**, *105*, 2563–2568. [CrossRef]
8. Long, C.; Wu, J.; Zhang, C.; Thomas, L.; Cheng, M.; Jenkins, N. Peer-to-peer energy trading in a community microgrid. In Proceedings of the 2017 IEEE Power & Energy Society General Meeting, Chicago, IL, USA, 16–20 July 2017; pp. 1–5.
9. Liu, T.; Tan, X.; Sun, B.; Wu, Y.; Guan, X.; Tsang, D.H. Energy management of cooperative microgrids with p2p energy sharing in distribution networks. In Proceedings of the 2015 IEEE international conference on smart grid communications (SmartGridComm), Miami, FL, USA, 2–5 November 2015; pp. 410–415.
10. Paudel, A.; Chaudhari, K.; Long, C.; Gooi, H.B. Peer-to-peer energy trading in a prosumer-based community microgrid: A game-theoretic model. *IEEE Trans. Ind. Electron.* **2018**, *66*, 6087–6097. [CrossRef]
11. Wang, N.; Xu, W.; Xu, Z.; Shao, W. Peer-to-peer energy trading among microgrids with multidimensional willingness. *Energies* **2018**, *11*, 3312. [CrossRef]
12. Sutton, R.S.; Barto, A.G. *Reinforcement Learning: An Introduction*; MIT press: Cambridge, MA, USA, 2018.
13. Chen, T.; Su, W. Indirect customer-to-customer energy trading with reinforcement learning. *IEEE Trans. Smart Grid* **2018**, *10*, 4338–4348. [CrossRef]
14. Chen, T.; Su, W. Local energy trading behavior modeling with deep reinforcement learning. *IEEE Access* **2018**, *6*, 62806–62814. [CrossRef]
15. Chen, T.; Bu, S. Realistic Peer-to-Peer Energy Trading Model for Microgrids using Deep Reinforcement Learning. In Proceedings of the 2019 IEEE PES Innovative Smart Grid Technologies Europe (ISGT-Europe), Bucharest, Romania, 29 September–2 October 2019; pp. 1–5.
16. Liu, Y.; Zhang, D.; Deng, C.; Wang, X. Deep Reinforcement Learning Approach for Autonomous Agents in Consumer-centric Electricity Market. In Proceedings of the 2020 5th IEEE International Conference on Big Data Analytics (ICBDA), Xiamen, China, 8–11 May 2020; pp. 37–41.

17. Başar, T.; Olsder, G.J. *Dynamic Noncooperative Game Theory*; SIAM: Philadelphia, PA, USA, 1998.
18. Chinchuluun, A.; Pardalos, P.; Migdalas, A.; Pitsoulis, L. *Pareto Optimality, Game Theory and Equilibria*; Springer: Berlin/Heidelberg, Germany, 2008; Volume 17. [CrossRef]
19. Alam, M.R.; St-Hilaire, M.; Kunz, T. Peer-to-peer energy trading among smart homes. *Appl. Energy* **2019**, *238*, 1434–1443. [CrossRef]
20. Blazev, A.S. *Global Energy Market Trends*; The Fairmont Press, Inc.: Lilburn, GA, USA, 2016.
21. Pazheri, F.; Othman, M.; Malik, N. A review on global renewable electricity scenario. *Renew. Sustain. Energy Rev.* **2014**, *31*, 835–845. [CrossRef]
22. Camus, C. Economic benefits of small PV "prosumers" in south European countries. In *EEIC2016 Scientific Session*; inesc-id: Lisbon, Portugal, 2016.
23. Lüth, A.; Zepter, J.M.; del Granado, P.C.; Egging, R. Local electricity market designs for peer-to-peer trading: The role of battery flexibility. *Appl. Energy* **2018**, *229*, 1233–1243. [CrossRef]
24. Myerson, R.B. *Game Theory*; Harvard University Press: Cambridge, MA, USA, 2013.
25. Branzei, R.; Dimitrov, D.; Tijs, S. *Models in Cooperative Game Theory*; Springer Science & Business Media: Berlin, Germany, 2008; Volume 556.
26. Fosler-Lussier, E. *Markov Models and Hidden Markov Models: A Brief Tutorial*; International Computer Science Institute: Berkeley, CA, USA, 1998.
27. Watkins, C.J.C.H. Learning from Delayed Rewards. Ph.D. Thesis, Psychology Department, University of Cambridge, Cambridge, UK, 1989.
28. Bellman, R. Dynamic programming. *Science* **1966**, *153*, 34–37. [CrossRef]
29. François-Lavet, V.; Henderson, P.; Islam, R.; Bellemare, M.G.; Pineau, J. An introduction to deep reinforcement learning. *arXiv* **2018**, arXiv:1811.12560.
30. Rummery, G.A.; Niranjan, M. *On-Line Q-Learning Using Connectionist Systems*; University of Cambridge, Department of Engineering: Cambridge, UK, 1994; Volume 37.
31. Sutton, R.S.; McAllester, D.A.; Singh, S.P.; Mansour, Y. Policy gradient methods for reinforcement learning with function approximation. In Proceedings of the Advances in Neural Information Processing Systems, Denver, CO, USA, 27–30 Nov 2000; pp. 1057–1063.
32. Watkins, C.J.; Dayan, P. Q-learning. *Mach. Learn.* **1992**, *8*, 279–292. [CrossRef]
33. Busoniu, L.; Babuska, R.; De Schutter, B.; Ernst, D. *Reinforcement Learning and Dynamic Programming Using Function Approximators*; CRC Press: Boca Raton, FL, USA, 2010; Volume 39.
34. Mnih, V.; Kavukcuoglu, K.; Silver, D.; Graves, A.; Antonoglou, I.; Wierstra, D.; Riedmiller, M. Playing atari with deep reinforcement learning. *arXiv* **2013**, arXiv:1312.5602.
35. Hochreiter, S.; Schmidhuber, J. Long short-term memory. *Neural Comput.* **1997**, *9*, 1735–1780. [CrossRef] [PubMed]
36. Xingjian, S.; Chen, Z.; Wang, H.; Yeung, D.Y.; Wong, W.K.; Woo, W.C. Convolutional LSTM network: A machine learning approach for precipitation nowcasting. In Proceedings of the Advances in Neural Information Processing Systems, Montreal, QC, Canada, 7–12 December 2015; pp. 802–810.
37. Nechchi, P. *Reinforcement Learning for Automated Trading*; Mathematical EngineeringPolitecnico di Milano: Milano, Italy, 2016.
38. Lucarelli, G.; Borrotti, M. A deep reinforcement learning approach for automated cryptocurrency trading. In Proceedings of the IFIP International Conference on Artificial Intelligence Applications and Innovations, Hersonissos, Crete, Greece, 24–26 May 2019; Springer: Berlin/Heidelberg, Germany, 2019; pp. 247–258.
39. Gao, G.; Wen, Y.; Wu, X.; Wang, R. Distributed Energy Trading and Scheduling among Microgrids via Multiagent Reinforcement Learning. *arXiv* **2020**, arXiv:2007.04517.
40. Kuate, R.T.; He, M.; Chli, M.; Wang, H.H. An intelligent broker agent for energy trading: An mdp approach. In Proceedings of the Twenty-Third International Joint Conference on Artificial Intelligence, Beijing, China, 3–9 August 2013.

Article

Distributed Generators Optimization Based on Multi-Objective Functions Using Manta Rays Foraging Optimization Algorithm (MRFO)

Mahmoud G. Hemeida [1,*], Salem Alkhalaf [2], Al-Attar A. Mohamed [3], Abdalla Ahmed Ibrahim [3] and Tomonobu Senjyu [4]

[1] Department of Electrical Engineering, Minia Higher Institute of Engineering, Minia 61111, Egypt
[2] Department of Computer Science, Arrass College of Science and Arts, Qassim University, Qassim 51431, Saudi Arabia; s.alkhalaf@qu.edu.sa
[3] Department of Electrical Engineering, Faculty of Engineering, Aswan University, Aswan 81511, Egypt; attar@aswu.edu.eg (A.-A.A.M.); draaibrahim@aswu.edu.eg (A.A.I.)
[4] Faculty of Engineering, University of the Ryukyus, 1 Senbaru, Nishihara-cho, Nakagami, Okinawa 903-70213, Japan; b985542@tec.u-ryukyu.ac.jp
* Correspondence: mahmod_hmeda2000@yahoo.com or mahmod_hmeda2000@aswu.edu.eg

Received: 16 June 2020; Accepted: 24 July 2020; Published: 27 July 2020

Abstract: Manta Ray Foraging Optimization Algorithm (MRFO) is a new bio-inspired, meta-heuristic algorithm. MRFO algorithm has been used for the first time to optimize a multi-objective problem. The best size and location of distributed generations (DG) units have been determined to optimize three different objective functions. Minimization of active power loss, minimization of voltage deviation, and maximization of voltage stability index has been achieved through optimizing DG units under different power factor values, unity, 0.95, 0.866, and optimum value. MRFO has been applied to optimize DGs integrated with two well-known radial distribution power systems: IEEE 33-bus and 69-bus systems. The simulation results have been compared to different optimization algorithms in different cases. The results provide clear evidence of the superiority of MRFO that defind before (Manta Ray Foraging Optimization Algorithm. Quasi-Oppositional Differential Evolution Lévy Flights Algorithm (QODELFA), Stochastic Fractal Search Algorithm (SFSA), Genetics Algorithm (GA), Comprehensive Teaching Learning-Based Optimization (CTLBO), Comprehensive Teaching Learning-Based Optimization (CTLBO (ε constraint)), Multi-Objective Harris Hawks Optimization (MOHHO), Multi-Objective Improved Harris Hawks Optimization (MOIHHO), Multi-Objective Particle Swarm Optimization (MOPSO), and Multi-Objective Particle Swarm Optimization (MOWOA) in terms of power loss, Voltage Stability Index (VSI), and voltage deviation for a wide range of operating conditions. It is clear that voltage buses are improved; and power losses are decreased in both IEEE 33-bus and IEEE 69-bus system for all studied cases. MRFO algorithm gives good results with a smaller number of iterations, which means saving the time required for solving the problem and saving energy. Using the new MRFO technique has a promising future in optimizing different power system problems.

Keywords: optimization techniques; manta ray foraging optimization algorithm; multi-objective function; radial networks; optimal power flow

1. Introduction

Distributed generation (DG) represents the production of electrical energy from distributed units near the consumers with a small capacity. DGs may be renewable or nonrenewable resources such as: wind turbine, solar Photo Voltaic (PV) geothermal, hydro, and diesel generators [1]. In the future, the shape of electrical power systems is expected to be one of four types. These types are centralized integrated with distributed generation, centralized with increased decentralization, partially decentralized, and fully decentralized. This shape will be formed based on many factors that may be summarized into three main categories. Firstly, high level of uncertainty that includes social factors, demand response, regulatory, policy and political factors; secondly, medium level of uncertainty that includes system challenges and technology requirements, and; finally, low level of uncertainty, which includes geographical, climatic considerations, availability of natural resources, population density and distribution, and existing infrastructure [2].

Attention towards DGs is increasing rapidly due to their effectiveness in solving overloading capacity problems and reducing the capital cost of the system. Furthermore, they require a small area, they are easy to construct within a small time and reduce greenhouse gases. Recently, DGs have gained much more interest due to the developed technologies of DG, restrictions applied for installing new transmission lines, increasing demand, liberalization of the electricity market, and concerns about greenhouse emissions [3]. Conventional energy resources such as thermal, hydro-power plants, have many limitations such as high investment costs for construction and transmission over a long distance; they are complex and inefficient. Distributed generations (DGs), which are decentralized power plants, are used to eliminate these limitations [4]. DGs can be categorized into two classifications: dispatchable DG, which includes all controllable generation types, such as gas turbine, combined heat and power, hydro, and biomass; and non-dispatchable DG, which includes generators characterized by uncertainty and whose generated power cannot be accurately predicted, such as wind turbines and photovoltaic cells [5].

Sizing and allocation of DG is an important issue to improve system efficiency, reliability, and power quality [6]. Allocating DG inappropriately increases system losses, and, consequently, increases total cost [7]. On the other hand, a properly sized and located DG provides many technological and economic benefits. For technological advantages, it could minimize line losses, improve power, minimize emissions of CO_2, increase security, provide better power standard, and prevent transmission and distribution congestion. For economic advantages, it saves the investment required to upgrade the services, minimizes maintenance and operation costs, reduces expenditure required for curing environmental damages, reduces fuel rate, and security costs for important loads. The use of an unsuitably placed or improperly sized DG has an adverse effect on system performance. Adding two or more DGs provides more advantages both economically and environmentally [8]. Renewable energy (RE) based DGs have been optimized for many objective functions such as: minimization of power losses, maximization of voltage stability, keeping stability margin, minimizing voltage deviation, minimizing total harmonic distortion, minimization of total costs, maximization of system reliability (minimizing system failures), maximizing RE penetration, and minimizing CO_2 emission. In order to assure system limits, some constraints are used. For instance, bus voltage limits, power flow equality constraint, overloading constraint, and DG capacity constraint [9].

A renewable energy (RE) based DG has been studied extensively during the last two decades. The effect of the Wind Turbine (WT)generator's uncertainty on DG has been studied using different algorithms [10]. The DG-based PV cell has been overviewed taking into account optimization techniques, constraints, and optimization techniques [11]. Hybrid combinations of WT and PV cells have been studied in order to minimize power loss and total cost and to reduce CO_2 emissions [12,13]. WT and PV generators have also been combined with other types of renewable and nonrenewable DGs such as: micro turbine, fuel cell, biomass, battery storage system, and capacitor bank in order to enhance power quality and to increase benefits of these hybrid combinations [14–17]. In real life, the multi-objective optimization problem has been applied to residential homes to maximize profits and minimize electricity charges by utilizing an energy storage system (ESS), heat pump, and an electric vehicle [18]. On the microgrid level, also, hybridization between PV, WT, ESS, and a fuel cell has been applied on a remote island in Japan in order to overcome the intermittent nature of a renewable energy system, reduce energy costs, and to improve system performance [19]. Due to load variance and the need to meet load demand, an ESS has been introduced as an effective solution to the optimal unit commitment problem to minimize operating costs, maximize profits, and to assure system reliability [20].

Optimization problems can be solved using single or multi-objective functions. Multi-objective problems may be solved using one of two methods. The first is to convert the multi-objective problem into a single-objective problem using weight factors which gives a single dependent solution. Any change in the parameter's weight, even if it is very small, causes a significant change in the prediction of the DG's size and location. The second method is to develop meta-heuristic algorithms in order to generate a set of Pareto-optimal solutions which are non-dominated solutions [21–23].

Many optimization techniques have been used to optimize the DG site and size. These techniques are categorized as follows: analytical approach, classical (non-heuristic) approach, meta-heuristic optimization approach, hybrid approach, and some other assorted approaches. The performance of the analytical and classical (non-heuristic) approach is effective for simple systems, but not for complex systems. Meta-heuristic algorithms offer fast convergence with high accuracy and suitability for complex and large systems. Hybrid optimization techniques provide better performance with higher accuracy and are more suited to complex multi-objective problems. Distribution Network (DN) contains uncertainties resulting from the variable output of DG, load variation, and energy cost. This produces a complex system that is not easy to optimize. Meta-heuristic and hybrid techniques provide a better solution for optimizing the allocation and capacity of DGs [6]. Meta-heuristic algorithms are a powerful tool in optimizing a wide range of problems. They are problem-independent algorithms. Meta-heuristic algorithms are categorized into two divisions: evolutionary algorithm (EA), and swarm intelligence algorithms (SI). The evolutionary algorithm contains many types of algorithms such as: genetic algorithm (GA), evolutionary strategy (ES), evolution programming immune algorithm (AIA), and Jaya algorithm (JA). Swarm intelligence algorithms represent the biggest category that contains particle swarm optimization (PSO), shuffled frog leaping optimization (SFLO), ant lion optimization algorithm (ALOA), ant colony optimization (ACO), firefly algorithm (FFA), grey wolf optimizer (GWO), dragon algorithm (DA), whale optimization algorithm (WOA), and social learning PSO (SLPSO) [24]. Meta-heuristic algorithms can also be classified into four main sections: methods based on an evolutionary algorithm, methods based on physics, methods based on swarm algorithms, and methods based on human activities [22].

Based on the type of objective function (single or multi-objective) and optimization techniques, DG integrated with DN has been extensively studied. Minimizing power losses as a single objective function has been studied widely through different optimization techniques and different constraints. Teaching-learning-based (TLBO), WOA, DA, and Moth-Flame Optimization Algorithm (MFO) optimization algorithms have been applied to allocating and sizing DG to minimize power loss [25,26]. Determination of the size and location of the DG and capacitor bank has been studied to minimize total power loss using a hybrid approach based on GA and Artificial Bee Colony Algorithm (ABC) [27]. The coyote algorithm (COA) has effectively minimized power losses through determining the size and location of DGs, the results have also been compared to that of many types of algorithms (Genetic Algorithm (GA), Particle Swarm Optimization (PSO), Cuckoo Search (CSA), Fireworks Algorithm (FWA), Stochastic Fractal Search (SFS), Harmony Search Algorithm (HSA), and Salp Swarm Algorithm (SSA)) [28]. The size and location of DG integrated with a 15 and 33 bus system have been optimized using the whale optimization algorithm (WOA) to enhance power loss reduction [29]. The Loss Sensitivity Factor (LSF)) and PSO have been used to minimize power loss using the exact loss' formula as a result of optimizing the size and location of the DG [30]. Compared to three types of optimization techniques (PSO, Improved Analytical (IA) method, hybrid approach), the genetic algorithm (GA) minimizes total power losses in 33 and 69 bus systems through optimizing the site and size of DG integrated with the two systems [31]. Allocation and sizing of DG connected to 33, 69, and 118-bus systems have been investigated to minimize power losses using chaotic stochastic fractal search

(CSFS) [32]. Differential evolution (DE) and moth swarm algorithms (MSA) have been successfully used to optimize distributed energy resource allocation to reduce power loss and voltage deviation [33].

Multi-objective non-dominating solution based algorithms have been used to solve optimization problems as an effective strategy. Different objective functions have been studied, many types of optimization techniques have been used, and various kinds of constraints and limitations have been applied in the optimization of DG connected to the grid. Power losses and cost functions have been simultaneously minimized by finding the optimum size and location of PV- and WT-based DG through applying the non-dominated sorting genetic algorithm II (NSGA-II) [34]. Sizing and allocation of multiple DGs with unity power factor and capacitor bank have been adjusted to optimize different objective functions. Three objective functions are optimized in parallel: total power loss, voltage deviation, and load-balancing using multi-objective (hybrid weight improved particle swarm optimization (WIPSO)-gravitational search algorithm (GSA)) or (WIPSO-GSA) [35]. The multi-objective particle swarm optimization (MOPSO) algorithm has been applied to optimize two different objective functions simultaneously: cost function, and the total power losses during failures [36]. Grid-based multi-objective harmony search algorithm (GrMHSA) has been used to determine the optimal location and capacity of DG. This is to minimize multi-objectives simultaneously, total voltage deviation, active and reactive power loss [37]. Active power loss and Fast Voltage Stability Index (FVSI)) have been simultaneously minimized using multi-objective chaotic mutation immune evolutionary programming (MOCMIEP) through determining the best size and location of distributed generation photovoltaic (DGPV) [38].

The voltage deviation at any bus is the difference between the source voltage (reference voltage) and node voltage at each bus. It would be preferred that it be minimized. The voltage deviation index is expected to be one of the most important indices for safe operation, power quality, and increasing demand; it may result gradually in deteriorating voltage stability; the voltage may collapse, and increasing node voltages may reduce reactive power losses in the system [39,40].

Voltage instability has a damaging impact on the power system which in some cases may lead to partial or total blackout. Voltage Stability Index (VSI)is an indication of the voltage level at any bus, that level must be maintained within acceptable limits. It must be always greater than zero. VSI is mainly used to determine the most sensitive buses which have the lowest VSI. The maximizing voltage stability index has been introduced as an objective function in optimizing the size and location of DG in many pieces of research. DG must be allocated at buses with the lowest VSI (most sensitive) in order to increase system stability at that point in the system [41,42].

Due to the importance of power system efficiency and fixed voltage level, minimizing power loss and improving system voltage is expected to be a major objective function. Many pieces of research added some objective functions besides power loss and voltage profile. The most known objective functions besides power loss and voltage profile are reactive power loss, voltage stability index, investment cost, operational cost, total harmonic distortion, and system load ability, which have been introduced as a weighted sum multi-objective function to determine best site and size of DG integrated with DN [43–50].

Not all optimization algorithms are suitable for solving multi-objective problems. The weight sum method is suitable to modify multi-objective functions into a single objective one. It makes matters easy for different types of algorithms to cope with multi-objective problems. This method provides a dominated solution based on pre-determined weight factors. The decision-maker should accurately determine weight factors to achieve good results. Many optimization techniques have been used to optimize size and location in order to improve voltage profile and minimize power loss. Applying three different objective functions such as minimizing power loss, maximizing voltage stability index, and minimizing voltage deviation provide good results in improving power quality and enhancing system efficiency. Many optimization techniques have been applied based on these three objective functions. These techniques are the quasi-oppositional differential evolution Lévy flights algorithm (QODELFA), a novel stochastic fractal search algorithm (SFSA), genetics algorithm (GA), a comprehensive teaching learning-based optimization (CTLBO) technique, ant-lion optimization algorithm (ALOA), improved Harris Hawks optimizer (IHHO), and multi-objective improved Harris Hawks (MOIHHO) [51–58].

Bio-inspired optimization algorithms are artificial intelligence algorithms based on the characteristics or behaviors of living creatures. They are mostly inspired by the behaviors of insects, birds, fish, or any group-living animals. Many complicated industrial and engineering problems have been solved using bio-inspired optimization techniques. They also have many features that can be extensively used. For instance, they are flexible, simple, easy to understand and implement; they have the ability to avoid trapping in local optima. Swarm intelligence (SI)-based algorithms are types of bio-inspired meta-heuristic algorithms that consider that a group of individuals (SWARM) provides a better solution compared to that of separate individuals. Compared to the evolutionary type algorithm, swarm intelligence has fewer operators, fewer parameters, memorization of best results, and ease of implementation. The manta ray foraging optimizer (MRFO) is one of the swarm intelligence (SI) based algorithms that imitate foraging behaviors - chain, cyclone, and somersault of manta rays [59].

MRFOA is a new powerful optimization algorithm. In this paper, the MRFO algorithm has been applied for the first time to optimize multi-objective problems. The multi-objective problem is converted into a single objective one using the weight sum method. Optimizing power loss, voltage deviation, and voltage stability index have been achieved through optimizing the size and location of DG units connected to the radial power system. Two radial distributed power systems are integrated with three DG units, 33-bus, and 69 bus systems. The efficiency of the proposed technique is investigated through optimizing DG units with different power factors. The rest of the paper is organized as follows: Section 2, the foraging strategies of manta ray, especially those simulated in the MRFO algorithm, are explained briefly. Then, the problem is formulated in Section 3, i.e., power flow analysis, objective function; and constraints are explained in Section 3. The mathematical model of MRFO is simulated in Section 4. The simulation results, the parameters of power systems, and the parameters of the optimization algorithm are described in Section 5. Finally, the conclusions and future work are listed in Section 6.

2. Manta Ray Foraging Strategies

Manta rays are one of the biggest deep-sea creatures. Their body is flat with two pectoral fins. They swim smoothly like birds in the sky. They have a huge terminal mouth behind two extended vertical lobes. Their main food is plankton which does not require sharp teeth. Manta rays and the main parts of their body are shown in Figure 1 [60]. Feeding strategies of the manta ray can be classified into eight types according to the number of rays and their swimming behavior. These strategies are: straight, surface, chain, piggy-back, somersault, cyclone, sideways, and bottom. They can also be divided into group and individual feeding. Individual feeding strategies are straight, surface, somersault, sideways, and bottom. The group foraging strategies are chain, cyclone, and piggy-back. Chain, somersault, and straight foraging are the dominant foraging behaviors at 40.7%, 29.47%, and 24.21% respectively. Piggyback foraging comes next with 2.46%, followed by cyclone foraging 2.11%. Finally, surface and sideways foraging have the lowest percentages at 0.7% and 0.35% [61]. The MRFO algorithm imitates the feeding attitude of manta rays that include chain, cyclone, and somersault foraging [60]. Only chain, cyclone, and somersault foraging strategies are going to be explained as a preamble to the MRFO algorithm.

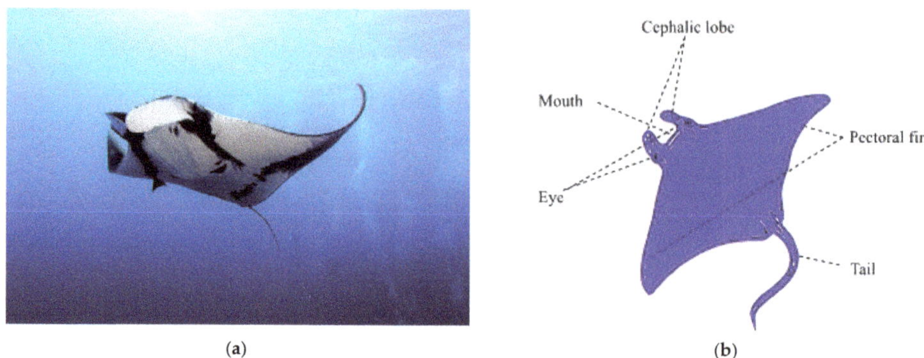

(a) (b)

Figure 1. Manta rays and the main parts of their body: (**a**) Manta ray; (**b**) Manta rays' physical construction.

2.1. Chain Feeding

A group of manta rays forms a line head-to-tail moving horizontally. They open their fins in front of their mouth and move forward and backward through the same area. Feeding runs sometimes extend to several hundred meters based on prey's concentration and distribution. At the end of feeding runs, they keep their line formed around prey and each of them updates its position slightly above or below the one in front. The group may expand to form a line of over 40 individuals at large feeding events. Six samples of manta rays' chain foraging are shown in Figure 2 [61].

Figure 2. Manta ray chain foraging strategy.

2.2. Cyclone Feeding

This foraging type is used when the prey is extremely concentrated in a limited area (plankton-rich water). Each individual in the mantas' line (feeding chain) circulates around itself until dragging prey into a large feeding circle. This circle's diameter is proportional to the number of animals joining the circle, approximately 15–20 m. The manta rays spiral motion lasts for a few minutes on average. The cyclone always rotates anticlockwise. The mantas' movement creates a vortex and the rotating movement of prey creates a current that draws prey outside the feeding circle towards them. The cyclone foraging technique is introduced in Figure 3 [61].

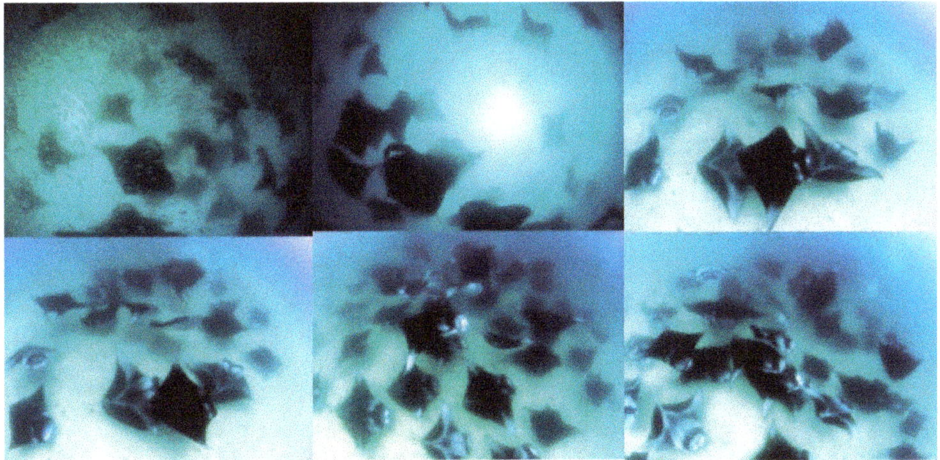

Figure 3. Cyclone foraging strategy.

2.3. Somersault Feeding

Somersault feeding is always backward. When the concentration of prey is high, the manta ray performs a backward feeding somersault completing a loop that has a diameter less than its body width. This type of feeding is usually performed when the prey is concentrated near the surface to restrict the prey's movement and increase feeding efficiency. During somersault feeding, they open their mouths widely and position their fins in front of their lower jaws. Figure 4 introduces different pictures that demonstrate the somersault foraging strategy [61].

Figure 4. Somersault foraging strategy.

3. Problem Formulation

Load flow analysis plays an important role in finding an accurate solution for system parameters. This will definitely affect finding the optimal solution for the DG installation problem [62]. The radial distribution network (RDN) has a radial structure that consists of a large number of buses and the ratio of *R/X* is high, which leads to a considerable voltage drop. Conventional methods such as Gauss Siedel,

Newton Raphson, and fast decoupled are not efficient or fast enough. They need a large number of iterations to provide good results [63,64]. Backward/Forward Sweep BFS load flow analysis is considered to be the most efficient method used to analyze RDN. This method is based on Kirchhoff's voltage law (KVL) and Kirchhoff's current law (KCL) [65]. Applying the BFS method includes three main steps. These three steps are the backward sweep, forward sweep, and nodal current analysis. This method is based on achieving convergence when the voltages maximum deviation is less than the tolerance error (0.000001) [66].

3.1. Backward Forward Sweep Power Flow Method

To calculate system parameters using the BFS method, three steps are used. Firstly, identify the different layers of the system as shown in Figure 5. Secondly, calculate the injected current in each phase. Then, apply backward sweep. Finally, apply forward sweep. The final step is repeated until achieving convergence and providing good results [67]. The single line diagram of radial DN is represented in Figure 6. The mathematical formulation is derived as follows:

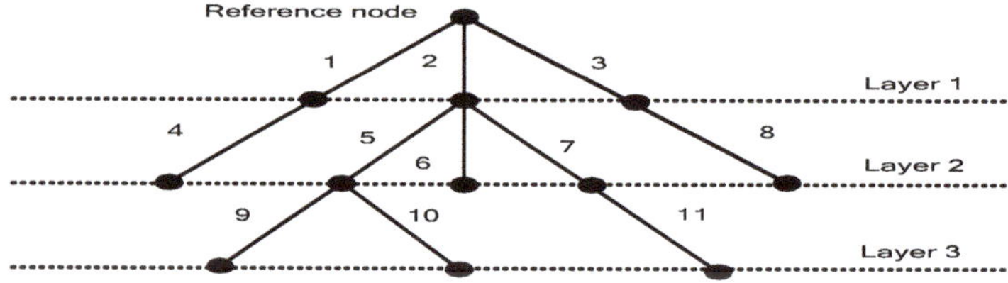

Figure 5. Layers in the radial distribution network (RDN).

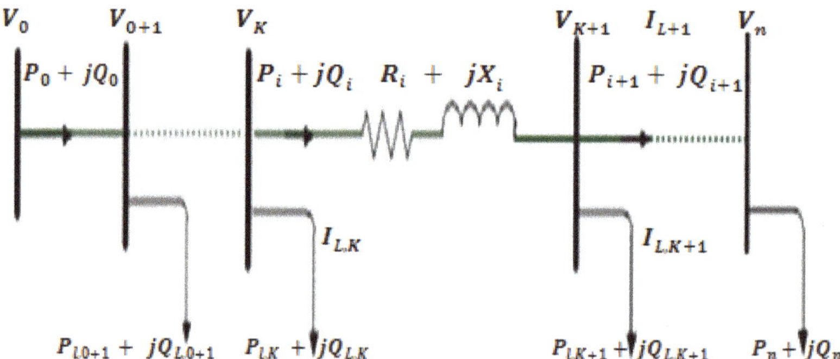

Figure 6. Single line diagram of RDN.

Step 1 define system 'layers' and calculate the load current of each bus I in the phase format depending on active and reactive power and the initial bus voltage using Equation (1).

Step 2 (backward sweep), the total branch currents are calculated starting from sub-lines and moving towards the main feeder which is represented by node #1, as given in Equation (2).

Step 3 (forward sweep), start to update bus' voltage from the main feeder at node #1 and move toward the last node using the mathematical formula shown in Equation (3) [65].

$$\overline{I_{L,k}} = \left(\frac{\overline{S_{L,k}}}{\overline{V_k}}\right)^* \tag{1}$$

$$\overline{I_L} = \overline{I_{L,k+1}} + \sum_{p\in M} \overline{I_{L+1}} \tag{2}$$

$$\overline{V_{k+1}} = \overline{V_k} - (R_l + jX_l)\overline{I_L} \tag{3}$$

$\overline{I_{L,k}}$ represents the load currents of each bus, $S_{L,k}$ represents system apparent power, V_i represents bus voltage, I_z is the total branch currents, and I_p is the sub-lines current.

3.2. Power Loss Calculation

The magnitudes of active and reactive powers are calculated based on Equations (4) and (5) respectively. To estimate the value of transmission line voltages, the mathematical formulation of Equation (6) can be used. Both active and reactive power losses are calculated for any line 1th using Equations (7) and (8) respectively. The total active power losses can be calculated from Equations (9). To simulate active and reactive power injection in the system at bus $k + 1$, Equations (10) and (11) are used [68].

$$P_{l+1} = P_l - P_{L,k+1} - R_l\left(\frac{P_l^2 + Q_l^2}{|V_k|^2}\right) \tag{4}$$

$$Q_{l+1} = Q_l - Q_{L,k+1} - X_l\left(\frac{P_l^2 + jQ_l^2}{|V_k|^2}\right) \tag{5}$$

$$V_{k+1}^2 = V_k^2 - 2(R_lP_l + X_lQ_l) + \left(R_l^2 + X_l^2\right)\left(\frac{P_l^2 + Q_l^2}{|V_k|^2}\right) \tag{6}$$

$$P_{loss(k,k+1)} = R_l\left(\frac{P_l^2 + Q_l^2}{|V_k|^2}\right) \tag{7}$$

$$Q_{loss(k,k+1)} = X_l\left(\frac{P_l^2 + Q_l^2}{|V_k|^2}\right) \tag{8}$$

$$P_{T\ loss} = \sum_{k=1}^{n-1} P_{loss(k,k+1)} \tag{9}$$

$$P_k = P_{k+1} + P_{L,k+1} + r_k\left(\frac{P_k^2 + jQ_k^2}{|V_k|^2}\right) - P_{DG} \tag{10}$$

$$Q_k = Q_{k+1} + Q_{L,k+1} + x_k\left(\frac{P_k^2 + jQ_k^2}{|V_k|^2}\right) - Q_{DG} \tag{11}$$

3.3. Objective Function

In order to optimize the size and location of DG, we have three different objective functions: minimization of power loss, minimization of voltage deviation, and maximization of voltage stability index.

3.3.1. Minimization of Total Active Power Loss

The total active power loss could be optimized using the following formula in Equation (12). PL is the total active power loss described by Equation (9), PL_b is the base power system losses [51].

$$OF_1 : \text{Min}\left(\frac{PL}{PL_b}\right) \tag{12}$$

3.3.2. Minimization of Voltage Deviation

Based on the following mathematical formula in Equation (13), voltage deviation is minimized. The voltage deviation denoted by VD is calculated using Equation (14). Base voltage deviation is denoted by VD_b, whereas V_i is the bus number, and V_r is the rated voltage (1.0 p.u.) [51].

$$OF_2 : \text{Min}\left(\frac{VD}{VD_b}\right) \tag{13}$$

$$\text{VD} = \sum_{i=1}^{n}(V_i - V_r)^2 \tag{14}$$

3.3.3. Maximization of the Voltage Stability Index

The voltage stability index is calculated using Equation (15). Then, the maximization of the voltage stability index (VSI) is turned into minimization using Equation (16). Finally, the objective function could be formulated using Equation (17) whereas VSI_b is the voltage stability index of the base case [51].

$$VSI_j = |V_i|^4 - 4\times\left(P_jX_k - Q_jR_k\right)^2 - 4\times\left(P_jR_k + Q_jX_k\right)\times|V_i|^2 \tag{15}$$

$$\text{Max}\left(VSI_j\right) = \text{Min}\left(\frac{1}{VSI_j}\right) \tag{16}$$

$$OF_3 : \text{Min}\left(\frac{VSI_b}{VSI_j}\right) \tag{17}$$

3.3.4. The Overall Objective Function

In order to use the MRFO algorithm to solve multi-objective problems, the single-objective functions are combined together using the weight sum method as shown in Equation (18) [51]. The weighting factors are taken as $\omega_1 = 1$, $\omega_2 = 0.65$, $\omega_3 = 0.35$ [51–53].

$$F = \text{Min}\ (\omega_1\times OF_1 + \omega_2\times OF_2 + \omega_3\times OF_3) \tag{18}$$

3.4. Operational Constraints

In order to simulate the system accurately, some constraints must be applied to the system. Three constraints must be achieved while optimizing our objective functions in this work described as follows:

3.4.1. Equality Constraints

The active and reactive power flow Equations (19) and (20) are used as equality constraints [51].

$$P_0 + \sum_{i=1}^{NDG} P_{DG}(i) = \sum_{i=1}^{k} P_{Li}(y) + \sum_{j=1}^{nb} P_{loss}(j) \tag{19}$$

$$Q_0 + \sum_{i=1}^{NDG} Q_{DG}(i) = \sum_{i=1}^{k} Q_{Li}(y) + \sum_{j=1}^{nb} Q_{loss}(j) \tag{20}$$

P_0 and Q_0 stand for the base active and reactive powers of the system respectively. $P_{Li}(y)$ and $Q_{Li}(y)$ are active and reactive powers supplied from the reference bus. NDG is the number of distributed integrated units.

3.4.2. Inequality Constraints

- Bus voltage constraints
  The magnitudes of voltage for all buses must be restricted in the range (0.95: 1.05)per unit, according to Equation (21) [51].

$$V_{min} \leq V_i \leq V_{max} \quad (21)$$

- DG sizing limits
  The distributed generators' active and reactive powers must be within limits according to Equations (22) and (23) [51].

$$P_{DG,min} \leq P_{DG,i} \leq P_{DG,max} \quad (22)$$

$$PF_{DG,min} \leq PF_{DG,i} \leq PF_{DG,max} \quad (23)$$

4. Mathematical Model of MRFO

The MRFO algorithm starts to position individuals randomly using the following Equation (24):

$$X_{i,j}(:) = Lb_{i,j} + r(:).\left(Ub_{i,j} - Lb_{i,j}\right) \forall i \in N_{pop} \&\& j \in N_{var} \quad (24)$$

The position of manta rays is expressed by $X_{i,j}(:)$; also, lower and higher boundaries are denoted by $Lb_{i,j}$ and $Ub_{i,j}$ respectively; while N_{pop} and N_{var} represent the number of populations and the number of variables respectively [69]. Manta rays have no sharp teeth; their main foodstuff is plankton which is a microscopic animal living in the water [60]. Manta rays have many foraging techniques. According to their swimming position, these techniques can be categorized into eight types. These types are: straight, surface, chain, piggy-back, somersault, cyclone, sideways, and bottom [61]. Only three strategies are simulated using the MRFO algorithm, which are chain, cyclone, and somersaulting foraging [60,69,70]. Chain and cyclone foraging are group-based techniques, but somersault foraging is an individual-based technique [61].

4.1. Chain Foraging

Plankton is not concentrated in one area. Manta rays look for their prey, i.e., plankton, and after locating its location, they swim directly towards it. Of course, the best position is that which comprises the highest plankton concentrations. Manta rays line up head-to-tail forming a chain. According to the prey's best position so far, manta rays change their position. Manta rays update their position following the mathematical formula of the next Equation (25) [60].

$$x_i^d(t+1) = \begin{cases} x_i^d(t) + r.\left(x_{Best}^d(t) - x_i^d(t)\right) + \alpha.\left(x_{Best}^d(t) - x_i^d(t)\right) & i = 1 \\ x_i^d(t) + r.\left(x^d{}_{i-1}(t) - x_i^d(t)\right) + \alpha.\left(x_{Best}^d(t) - x_i^d(t)\right) & i = 2, \ldots\ldots\ldots, N \end{cases} \quad (25)$$

$$\alpha = 2.r.\sqrt{|\log(r)|} \quad (26)$$

The updated position is expressed by $x^d{}_{i-1}(t)$. Iteration number and dimension are expressed by t and d respectively. $x_i^d(t)$ is the current position of ith individual; r is a random vector extended in the range of [0–1]. The weight coefficient is denoted by α; the best position is expressed by $x_{Best}^d(t)$.

4.2. Cyclone Foraging

The movement of each manta ray is restricted by two factors: the position of food, and the position of the one in front. Manta rays move towards plankton taking a spiral motion. This spiral-shaped movement can be simulated following a mathematical formula represented in Equation (27) [60].

$$\begin{cases} X_i(t+1) = X_{best} + r \times (X_{i-1}(t) - X_i(t)) + e^{b\omega} \times \cos(2\pi\omega) \times (X_{best} - X_i(t)) \\ Y_i(t+1) = Y_{best} + r \times (Y_{i-1}(t) - Y_i(t)) + e^{b\omega} \times \sin(2\pi\omega) \times (Y_{best} - Y_i(t)) \end{cases} \tag{27}$$

where ω represents a random number extended along the range of [0–1]. This motion is extended to an *n-D* space formula which simulates cyclone foraging. After this modification, the cyclone foraging can be expressed using the mathematical formula in Equation (28) [60].

$$x_i^d(t+1) = \begin{cases} x_{best}^d + r \times \left(x_{best}^d(t) - x_i^d(t)\right) + \beta \times \left(x_{best}^d(t) - x_i^d(t)\right) & i = 1 \\ x_{best}^d + r \times \left(x^d{}_{i-1}(t) - x_i^d(t)\right) + \beta \times \left(x_{best}^d(t) - x_i^d(t)\right) & i = 2, \ldots\ldots., N \end{cases} \tag{28}$$

$$\beta = 2e^{r_1 \frac{Tmax - t + 1}{T}} \times \sin(2\pi r_1) \tag{29}$$

where β stands for a weight coefficient; T_{max} is the maximum number of iterations; r_1 is the random number in the interval [0–1] The cyclone foraging improves both exploration and exploitation because each individual updates its position relying on the food position which varies randomly. A new random position is introduced as a new reference. As a result, the global best position is improved. This part of the optimization technique can be expressed using Equation (31), where the random position is represented by x_{rand}^d [60].

$$x_{rand}^d = Lb^d + r \times \left(Ub^d - Lb^d\right) \tag{30}$$

$$x_i^d(t+1) = \begin{cases} x_{rand}^d + r \times \left(x_{rand}^d(t) - x_i^d(t)\right) + \beta \times \left(x_{rand}^d(t) - x_i^d(t)\right) & i = 1 \\ x_{rand}^d + r \times \left(x^d{}_{i-1}(t) - x_i^d(t)\right) + \beta \times \left(x_{rand}^d(t) - x_i^d(t)\right) & i = 2, \ldots\ldots., N \end{cases} \tag{31}$$

4.3. Somersault Foraging

Somersault is an individual feeding strategy. Each individual moves toward the planktons' position then somersaults to a new position. Manta rays always swim around a higher food concentration area and update its position accordingly. Somersault foraging strategy can be derived from mathematical Equation (32) [60].

$$x_i^d(t+1) = x_i^d(t) + S \times \left(r_2 \times x_{best}^d - r_3 \times x_i^d(t)\right), i = 1, \ldots\ldots\ldots\ldots N \tag{32}$$

where, the somersault factor is expressed by S which decides the somersault range. Both r_2 and r_3 are random numbers extended in the range of [0–1]. Each individual is free to swim between its current position and the global best position determined so far. After some iteration, all individuals come closer to the optimum solution in the search space. As a result, increasing the number of iterations will decrease the range of somersault (inversely proportional) [60].

4.4. General Procedures of the MRFO Approach

The MRFO flow chart is demonstrated in Figure 7 and can be explained through the following steps [60,69].

1 - Formulating the optimization problem and determining boundary limits.
2 - Inserting control parameters, number of iterations (T_{max}), number of populations (N_{pop}), and somersault factor (S).
3 - Initially positioning individuals randomly and calculating the fitness of each to determine the best solution so far.
4 - Starting the main loop for i = 1: N_{pop}, If the stop criteria is not satisfied.
5 - If Rand is >0.5, then apply cyclone foraging.

- If t/T_{max} is < Rand, then update location using Equations (31)
- Else update location using Equation (28)
- End if

6 - Else (if Rand is <0.5) apply chain foraging.

- Update location using Equation (25)
- End if

7 - Evaluating the fitness value of each individual and updating position according to the best position.
8 - Then, update location using somersault foraging Equation (32)

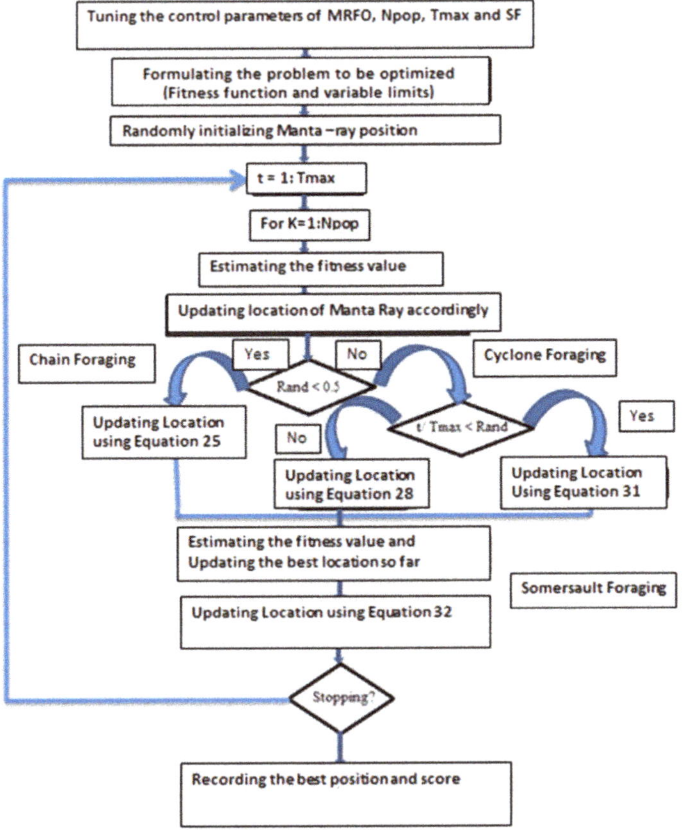

Figure 7. Manta Ray Foraging Optimization (MRFO) flow chart.

5. Results and Discussion

The MRFO algorithm has been applied to optimize three DG units integrated with two test systems: the IEEE-33 bus, and 69-bus systems. The power flow analysis is achieved using the backward-forward sweep method (BFS). The two systems are optimized based on three objective functions, minimizing APL, VD, and VSI^{-1}. DG units could be categorized based on the capability of injecting active and reactive powers, in other words, (operating power factor). Some types could provide only an active power such as photovoltaic and fuel cells; that means that they are operating at unity power factor. DG operating based on synchronous machines delivers both active and reactive power, and they are operated at non-unity power factor [51].

To investigate the performance of the proposed algorithm, four different cases have been studied for each system based on the operating power factor of the DG units. These four cases are unity, 0.95, 0.866, and optimum power factors operating conditions. For each case, the resulting power loss at each branch, the voltage magnitude at each bus m, and the voltage stability index (VSI) at each branch have been compared to that of the base case (without connecting DGs). The objective function is simulated using weight sum method with the following factors $\omega_1 = 1$, $\omega_2 = 0.65$, $\omega_3 = 0.35$. The parameters of MRFO are defined as follows: the MRFO algorithm was simulated at 50 search agents and 50 maximum iteration numbers and the somersault factor $(S) = 2$. The system was simulated using MATLAB 2014b running on a system core i7, 2.7GHZ, 4.0 GB ram. The studied cases are divided into two cases based on the power system and each case divided into four sub-cases based on different power factor operating conditions.

5.1. IEEE 33-Bus System

IEEE 33 bus-system is a standard electrical power system. It has a total load demand of 3.715 MW and 2.300 MVAR at 12.6 KV [54]. In this case, we have five subcases:

- Three DG units operating at a unity power factor
- Three DG units operating at 0.95 power factor
- Three DG units operating at 0.866 power factor
- Three DG units operating at an optimum power factor
- Comparing results for different power factors

5.1.1. Three DG Units at Unity Power Factor

The simulation results of power loss, voltage magnitudes, and VSI for the whole system are compared to the results of the base case and represented by Figures 8–10, respectively. Looking at Figure 8, the power loss is decreased significantly around every branch. The maximum power loss recorded at bus 2 equals 52.0726 and is minimized to be 14.0135; the minimum power loss recorded at branch 2 equals 0.0132 and is minimized to be 0.0115. The voltage at bus 18 was the minimum value of 0.90378 and is maximized to be 0.98; the maximum voltage at bus 2 is maximized to be 0.99909 as shown in Figure 9. The VSI is simulated as shown in Figure 10. The minimum value recorded at branch 17 equals 0.6672 and is enhanced to be 0.9182; the maximum VSI at branch 1 equals 0.9881 and is improved to be 0.9963.

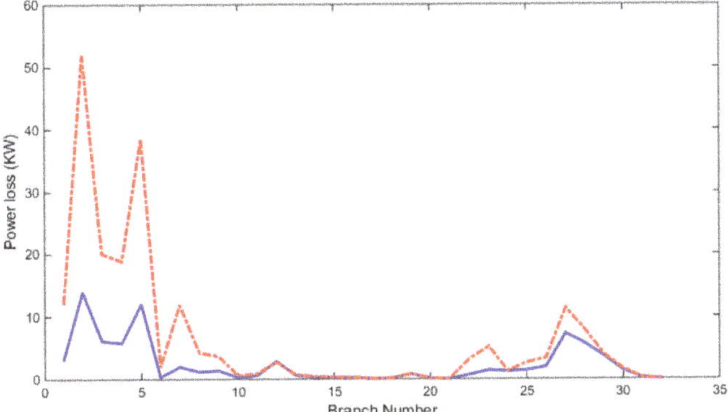

Figure 8. Characteristics of power loss at unity pf.

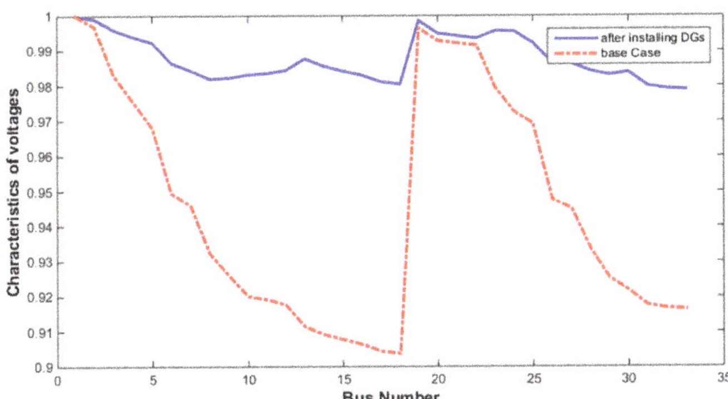

Figure 9. Characteristics of voltage at unity pf.

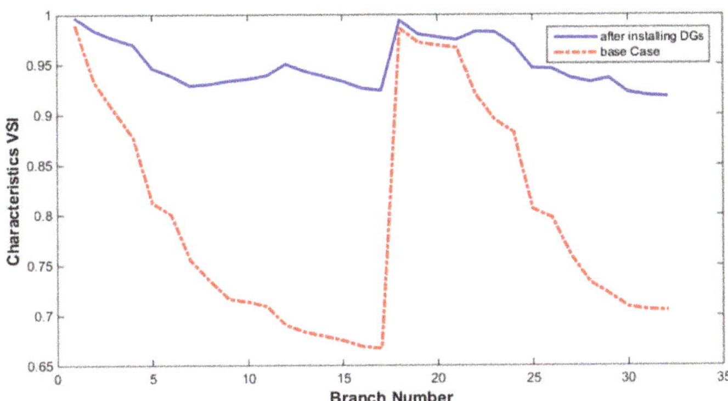

Figure 10. Characteristics of VSI at unity pf.

Table 1 shows the results obtained by the MRFO algorithm to allocating and sizing three units of DGs operating at the unity power factor. Additionally, a comparison is carried out between those results and the results obtained by nine different recent optimization techniques [51–57]. Three objective functions are considered: power loss minimization, voltage deviation minimization, and voltage stability index maximization. The weight factors are 1, 0.65, and 0.35 respectively [51,52].

The MRFO algorithm improves the overall system performance and provides better results compared to those obtained by other techniques. The optimum bus locations for DG units obtained by the MRFO algorithm are 30, 24, and 13; the same results are obtained by QODELFA, SFSA, and CTLBO (ε constraint) [51,52,54]. The active power loss is minimized to 77.3793 kW with a power loss reduction equal to 63.32%. The obtained result for power reduction is better than that obtained from all other techniques. In addition, the voltage deviation is minimized to be 0.0063 more than the lowest value obtained by the GA with 0.0056, which is a very small value [53]. The voltage stability index (VSI) is maximized to 0.9182 which is the same for QODELFA and SFSA, but it is a lower value compared to other techniques. The highest VSI obtained by the GA and CTLBO (ε constraint) in [53,54], but with the highest power loss values of 95.8 and 96.17 respectively.

Table 1. 33 bus system, three DG units, at unity Power factor (pf)

Reference	Method	Location	Size kW	APL kW	VD	VSI	LR (%)
Proposed	MRFO	30 24 13	1302.5 1136.4 962.292	77.3793	0.0063	0.9182	63.3239
[51]	QODELFA	13 24 30	964.7 1133.4 1301.7	77.408	0.00621	0.9182	63.31
[52]	SFSA	13 24 30	964.7 1133.7 1301.8	77.410	0.006232	0.9182	63.31
[53]	GA	25 30 13	909.0 1684.0 1658.0	95.8	0.0007	0.9701	54.6
[54]	CTLBO	13 25 30	1036.4 1163.1 1521.7	85.9595	0.0026	0.9481	59.26
[54]	CTLBO ε constraint	13 24 30	1.1926 0.8706 1.6296	96.1732	0.0009	0.9638	54.4
[56]	MOHHO	13 25 31	1207 763 1400	92.95	0.002	0.9654	55.94
[56]	MOIHHO	14 24 31	1223 1144 1290	92.25	0.0019	0.9580	56.27
[57]	MOPSO	12 25 33	1200 949.8 1142.7	83.99	0.0053	0.919	60.19
[57]	MOWOA	14 24 31	1021.6 1200 1200	79.72	0.0045	0.9249	62.214

5.1.2. Three DG Units at 0.95 Power Factor

Three basic characteristics, namely power loss, bus voltage, and VSI, are simulated as shown in Figures 11–13 as a comparison between DGs operating at 0.95 power factor and the base case. This is to investigate the effect of installing DG units and the validity of the MRFO algorithm. Figure 11 represents the system power loss. The maximum power loss recorded at bus 2 is reduced to be 3.1636 kW; the minimum power loss at branch 32 is decreased to be 0.0113. The minimum voltage magnitude evaluated at bus 18 is maximized to be 0.9943 as shown in Figure 12. The minimum VSI recorded at bus 17 is maximized to 0.9773 as shown in Figure 13. The provided results are better compared to the previous case.

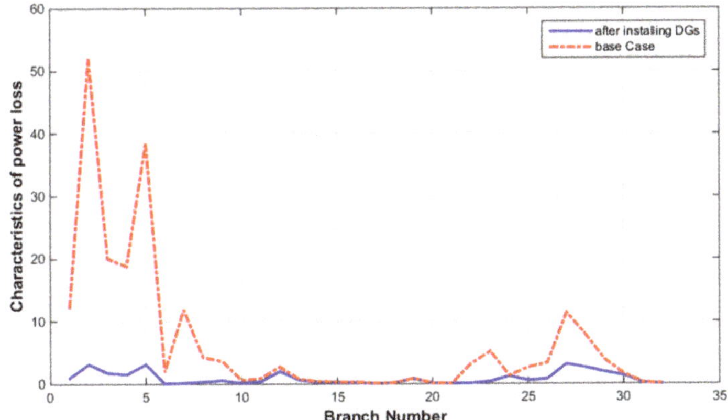

Figure 11. Characteristics of power loss, 33-bus system, 0.95 pf.

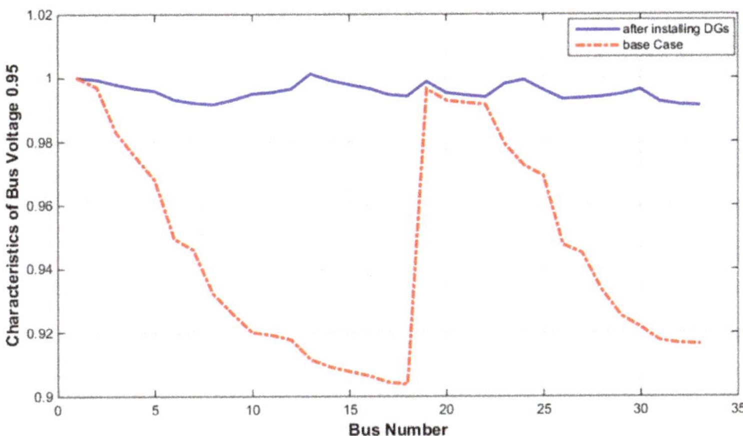

Figure 12. Characteristics of bus voltage, 33 bus system 0.95 pf.

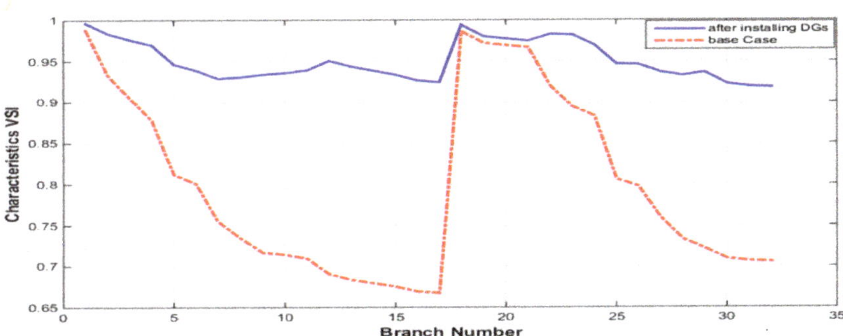

Figure 13. Characteristics of VSI, 33-bus system at 0.95 pf.

Optimization of three DG units operating at 0.95 lagging power factor based on the MRFO algorithm; and the data obtained by different techniques are shown in Table 2. The MRFO algorithm determines the best location for DG units, namely 30, 24, and 13, which are similar to results obtained by QODELFA, SFSA, and MOIHHO techniques [51,52,56]. The MRFO algorithm minimizes the total power loss to 29.13 with a total loss reduction of 86.19% which is the best value compared to all other techniques. The VSI provided by the proposed techniques is 0.966; it is lower than the highest value obtained by MOIHHO [56] with approximately 0.0128 which is a very small value. The voltage deviation optimized based on MRFO is minimized to 0.0008, which is higher than the best deviation obtained by MOIHHO with 0.0004. The overall performance is better than the results provided by the previous case.

Table 2. 33-bus, three DG units at 0.95 lag pf.

Reference	Method	Location	Size (kW)	Size (KVAR)	APL (kW)	VD	VSI	LR (%)
proposed	MRFO	30 24 13	1297.20 1098.00 911.31	426.36 360.88 299.53	29.1317	0.0008	0.9662	86.1922
[51]	QODELFA	13 24 30	916.9 1146.6 1316.7	301.3 376.8 432.7	29.386	0.0007	0.9698	86.07
[52]	SFSA	13 24 30	917.4 1146.3 1315.7	301.5 376.8 432.4	29.383	0.000673	0.9697	86.07
[56]	MOHHO	13 25 30	1008.0 910.0 1334.0	331.0 299.0 439.0	31.4	0.0005	0.976	85.1171
[56]	MOIHHO	13 24 30	924 1312 1356	304 431 446	30.6	0.0004	0.979	85.4963

5.1.3. Three DG Units at 0.866 Power Factor

Compared to the base case, three DG units are optimized and the simulation results of power loss, voltage magnitude, and VSI are shown in Figures 14–16. The maximum power loss at branch 2 is minimized to 0.4296 kW as represented in Figure 14. The minimum voltage recorded at bus 18 is maximized to 0.9964 as shown in Figure 15. The minimum voltage stability index recorded at bus 17 is improved to 0.9856 as described in Figure 16. This case provides results better than in the previous two cases.

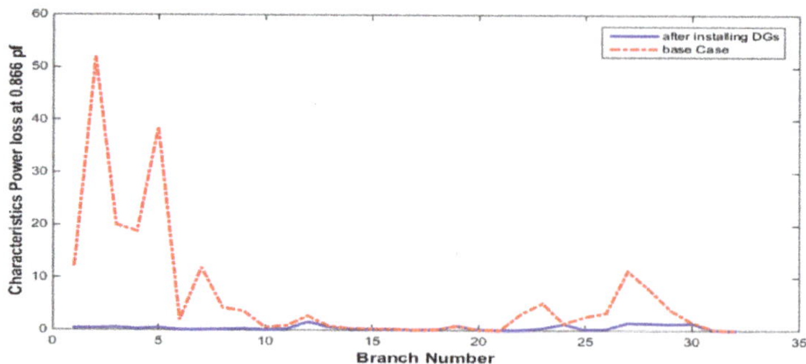

Figure 14. Characteristics of power loss, 33-bus system at 0.866 pf.

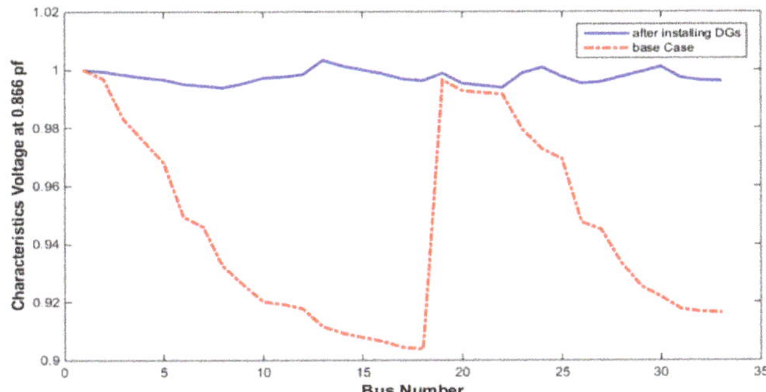

Figure 15. Characteristics of bus voltages, 33-bus system at 0.866 pf.

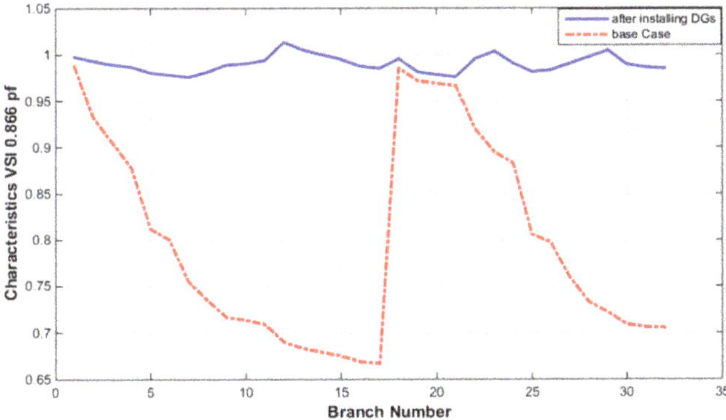

Figure 16. Characteristics of VSI, 33-bus system 0.866pf.

Table 3 lists the data obtained for optimizing three DG units operating at 0.866 lagging power factor. MRFO algorithm is compared to the QODELFA algorithm [51]. MRFO determines buses 13, 24, and 30 as the best location similar to results obtained by QODELFA. MRFO could not provide a significant improvement. But MRFO only takes 50 iterations to achieve optimum results, which is one-fourth of the number of iterations taken by QODELFA. The obtained results are almost the same for the two algorithms. The overall performance is improved compared to the former cases.

Table 3. 33-bus, three distributed generator (DG) units, 0.866 pf.

Reference	Method	Location	Size		APL (kW)	VD	VSI	LR (%)
			(kW)	(KVAR)				
proposed	MRFO	13	792.710	457.72	15.4956	0.00035	0.9763	92.6554
		24	1039.7	600.35				
		30	1239.6	715.75				
[51]	QODELFA	13	791.1	456.7	15.498	0.0003	0.9764	92.65
		24	1041.1	599.1				
		30	1243.1	717.8				

5.1.4. Three DG Units at Optimum Power Factor

Unlike the fixed power factor operation in the previous cases, the three DGs units optimized are operating at an optimum power factor which is not fixed. Compared to the former cases, this one gives the best results. As shown in Figure 17, the maximum power loss at bus 2 is minimized to a very low value of 0.3347 kW. The minimum voltage recorded at bus 18 is maximized to 0.9960 as demonstrated in Figure 18. Finally, the VSI which records its minimum value at bus 17, is maximized to a higher value of 0.9839 as shown in Figure 19.

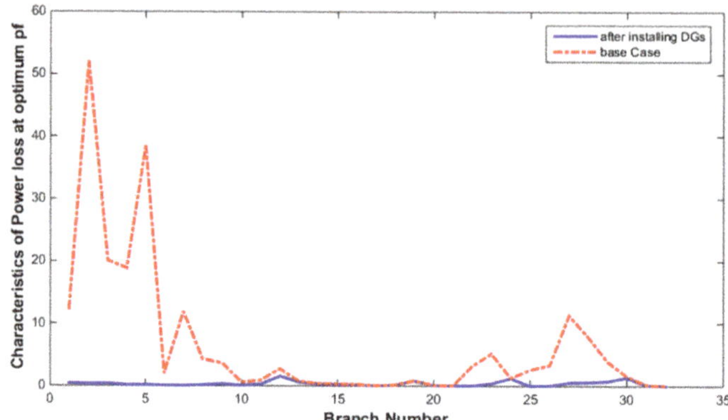

Figure 17. Characteristics of power loss, 33-bus system at optimum pf.

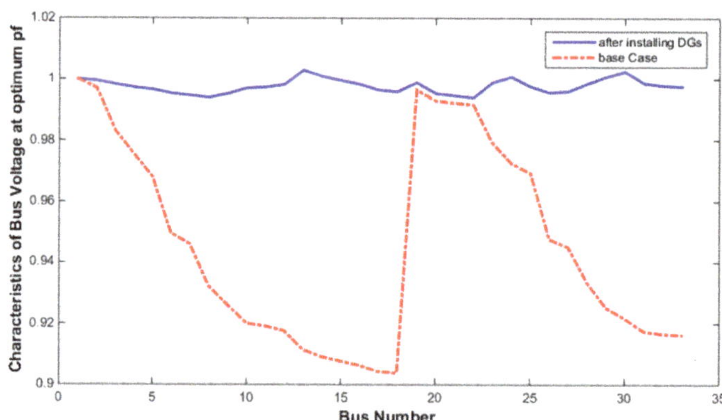

Figure 18. Characteristics of bus voltage, 33-bus at optimum pf.

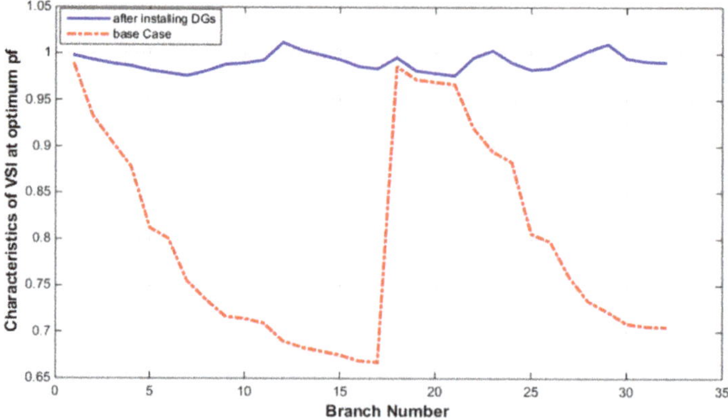

Figure 19. Characteristics of VSI, 33-bus system at optimum pf.

The MRFO algorithm is used to optimize DG units' size and location based on three different objective functions, namely APL, VD, and VSI; the obtained results are compared to those of the other three techniques, namely SFSA, MOHHO, and MOIHHO [52,56] as listed in Table 4. Buses 13, 30, and 24 were determined to be the best locations to install DG units operating at optimum power factor. The same locations were obtained by SFSA [36]. The optimum power factors obtained by the MRFO algorithm are 0.8916, 0.7258, and 0.8963. The reduced power factor values mean that much more reactive power is injected into the system. As a result, the simulated characteristics are highly improved. Compared to the previous three cases, this one gives the best performance. The results obtained by MRFO and SFSA are almost the same and provide better results compared to the other two techniques, namely MOHHO and MOIHHO [56]. Although SFSA provides better loss reduction compared to our MRFO-almost 0.5%, MRFO takes only 50 iterations to get optimum results, which is half the number of iterations taken by SFSA.

Table 4. 33-bus, three DG units at optimum pf.

Reference	Method	Location	DG Capacity and PF			APL (kW)	VD	VSI	LR (%)
			(kW)	(KVAR)	(Pf)				
Proposed	MRFO	13 30 24	809.9112 1072.4 1079.0	411.27 1016.2 533.9	0.8916 0.7258 0.8963	11.918	0.000338	0.9760	93.8559
[52]	SFSA	13 24 30	834.0 1064.8 1059.2	391.6 531.4 1025.9	0.905 0.895 0.718	11.911	0.000334	0.9763	94.35
[56]	MOHHO	12 25 30	951 786 1381	516 436 809	0.88 0.87 0.86	18.8	0.0005	0.978	91.0892
[56]	MOIHHO	12 24 30	916 1088 1171	576 386 830	0.85 0.94 0.82	15.0	0.0003	0.978	92.8903

5.1.5. Comparing Results Obtained for Different Power Factors

Table 5 is considered to be a collection of the previous four cases to investigate the effect of operating DG units at various power factor values. For the first objective function, the total active power loss reaches its minimum value of 11.918 with a total reduction of 93.86% when the DG units were operated at an optimum power factor. In addition, the second objective function, the voltage deviation, was minimized to 0.000338 as the lowest value for all four cases when the DG units were set to an optimum power factor. Finally, the voltage stability index was maximized to 0.976 as the highest value obtained by the optimum and 0.866 pf. SFSA [52] achieves results better than the MRFO algorithm; the total loss reduction is about 0.49% more than the loss reduction achieved by MRFO. Decreasing the operating power factor from unity to the optimum value improves the performance of the power system significantly due to increasing the injected reactive power suitably with the system conditions and the objective functions needed to be optimized.

Table 5. Comparing results for different power factors.

Case	Location	DG Capacity and PF			APL (kW)	VD	VSI	LR (%)
		(kW)	(KVAR)	(Pf)				
Optimum pf	13 30 24	809.9112 1072.4 1079.0	411.27 1016.2 533.9	0.8916 0.7258 0.8963	11.918	0.000338	0.976	93.8559
0.866 pf	13 24 30	792.710 1039.7 1239.6	457.72 600.35 715.75	0.866 0.866 0.866	15.4956	0.00035	0.976	92.6554
0.95 pf	30 24 13	1297.20 1098.00 911.31	426.36 360.88 299.53	0.95 0.95 0.95	29.1317	0.0008	0.9662	86.1922
Unity pf	30 24 13	1302.5 1136.4 962.292	0 0 0	1 1 1	77.3793	0.0063	0.9182	63.3239

5.2. IEEE 69-Bus System

IEEE 69 bus-system is an electrical standard power system. It has a total load demand of 3.8 MW and 2.69 MVAR at 12.6 KV [54]. In this case, we have five subcases:

- Three DG units operating at a unity power factor
- Three DG units operating at 0.95 power factor
- Three DG units operating at 0.82 power factor
- Three DG units operating at an optimum power factor
- Comparing results for different power factors

5.2.1. Three DG Units at Unity Power Factor

According to the objective functions, three characteristics of the power system are described, namely power loss, voltage profile, and voltage stability index (VSI). The MRFO algorithm optimizes the system to improve system performance. Power loss is shown in Figure 20; the maximum power loss recorded at branch 56 equals 49.6782 kW and is minimized to 14.8851 kW. The voltage characteristic is shown in Figure 21; the minimum voltage at bus 65 equals 0.90919 and is maximized to 0.98785. The minimum voltage stability index recorded at bus 64 equals 0.6833 and is maximized to 0.9522 as shown in Figure 22.

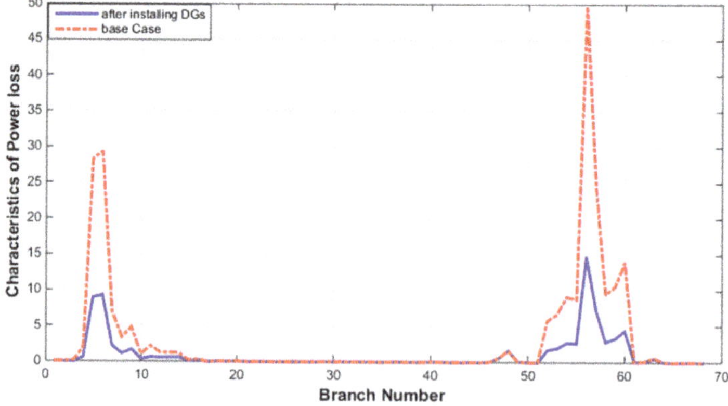

Figure 20. Characteristics of power loss, three DGs at unity pf.

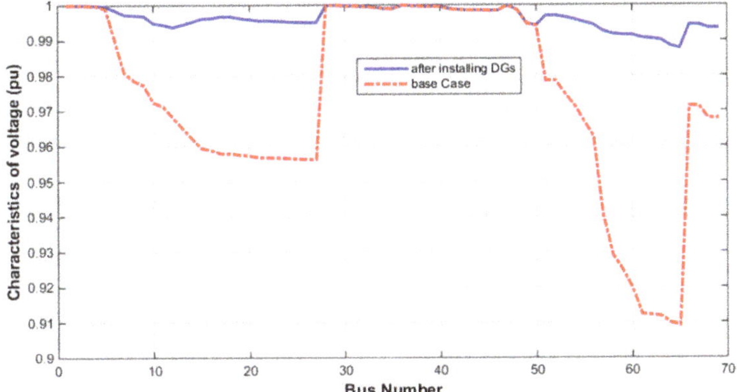

Figure 21. Characteristics of voltage, three DGs at unity pf.

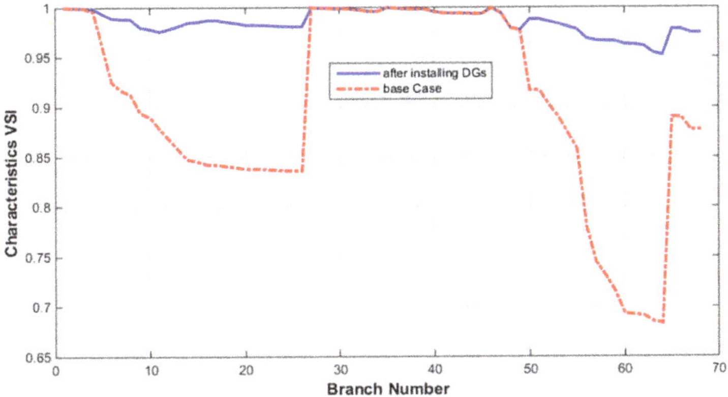

Figure 22. Characteristics of VSI, three DGs at unity pf.

MRFOA and eight other algorithms are compared to each other in optimizing three DG units based on the pre-determined objective functions, APL, VD, and VSI for the IEEE 69-bus system as shown in Table 6. The MRFO algorithm determines the optimum locations to be 19, 11, and 61 respectively; the same as SFSA [52]. The total active power loss is minimized to 71.02 with loss reduction equal to 68.427%, which is better than the results of all other techniques. Although the voltage deviation is minimized to 0.0022, it is not the minimum deviation value. It is 0.0019 higher than the minimum value reported by CTLPO (ε constraint), [54] but the second method estimated the total capacity for DG units as 3329.4 KW which is higher than that provided by MRFO, 2923.7 with approximately 405.7 kW. The voltage stability index is maximized to 0.94 lower than the best values obtained by CTLPO and CTLPO (ε constraint) 0.977, but those techniques provide higher power loss,76 kW, 37 kW, and 79.66 kW respectively, and that makes sense.

Table 6. 69-bus system, three DG units at unity Power Factor (PF)

Reference	Method	Location	Size (kW)	APL (kW)	VD	VSI	LR (%)
Proposed	MRFO	19 11 61	473.1375 591.3010 1859.3	71.0207	0.0022	0.9402	68.4268
[51]	QODELFA	11 20 61	629.4 438.6 1953.7	72.295	0.00150	0.9525	67.87
[52]	SFSA	11 19 61	570.3 466.1 1967.4	72.445	0.001434	0.9537	67.80
[54]	CTLPO	11 18 61	560.3 427.4 2153.4	76.372	0.0008	0.9770	66.0478
[54]	CTLPO ε constraint	12 25 61	965.8 230.7 2133.6	79.66	0.0003	0.9770	64.5861
[56]	MOHHO	20 60 61	643.6 971.4 1328.2	81.0	0.0008	0.9720	63.9904
[56]	MOIHHO	18 61 64	796.2 1447.1 707.5	80.8	0.0007	0.978	64.0793
[57]	MOPSO	21 61 66	383.6 1770.8 1450.2	82.79	0.0015	0.9455	63.1946
[57]	MOWOA	20 53 61	558.8 784.9 1970.1	75.56	0.00083	0.9643	66.4088

5.2.2. Three DG Units at 0.95 Power Factor

MRFO algorithm is applied to optimize three DG units operating at 0.95 power factor to optimize the proposed objective functions. The power loss is optimized as shown in Figure 23, i.e., the maximum power loss at branch 56 is reduced to 3.7085 kW. The minimum voltage magnitude recorded at bus 65 is maximized to 0.9963 as shown in Figure 24. The minimum value of VSI at branch 64 is maximized to 0.9855 as demonstrated by Figure 25. The obtained results are better compared to the previous case.

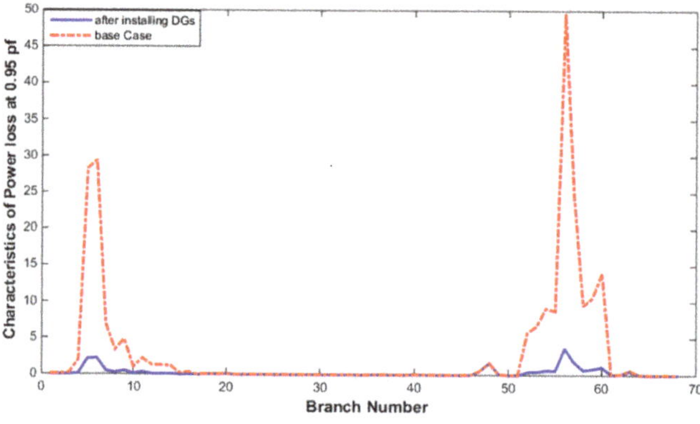

Figure 23. Characteristics of power loss at 0.95 pf.

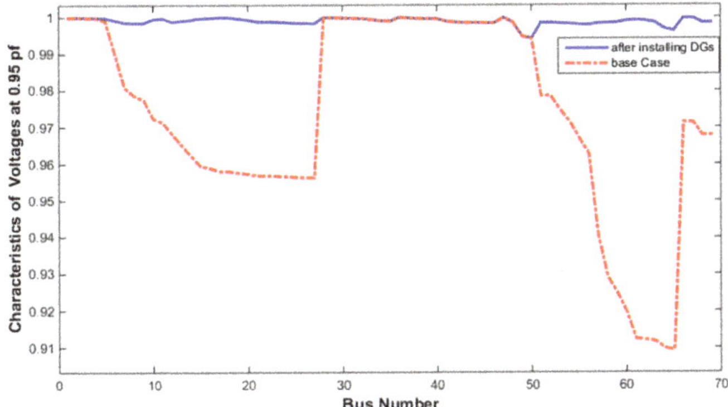

Figure 24. Characteristics of voltage at 0.95 pf.

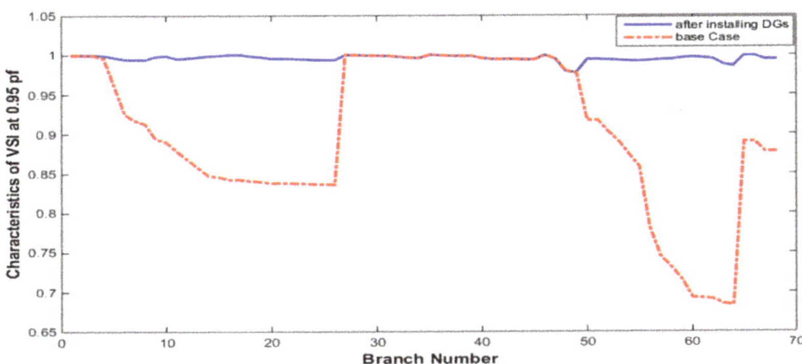

Figure 25. Characteristics of VSI, 0.95.

The IEEE 69-bus system is integrated with three units of DG. These DG units are optimized to improve system performance using the MRFO algorithm. The obtained results are compared to that of four other techniques and the data is recorded in Table 7. The selected buses to install DG units are 11, 18, and 61; this is similar to results obtained by QODELFA [51]. The results obtained by the MRFO algorithm and QODELFA are almost the same. The total power loss is minimized to 20.7702 kW which is the minimum value compared to other techniques. The total installed capacity 2919 kW and 959.65 KVAR close to QODELFA, 2915 kW, and 958 KVAR [51]. The voltage deviation is minimized to a value slightly higher than that for QODELFA with about 0.00001. The voltage stability index is maximized to 0.9772 which is the best value of all. The MRFO algorithm could not provide a significant improvement compared to the QODELF technique. However, MRFO is better as it takes one-fourth of the number of iterations taken by QODELF to achieve optimum results.

The obtained results are better compared to the previous case (unify power factor).

Table 7. 69-bus, three DG units at 0.95 lag pf.

Reference	Method	Location	Size (kW)	Size (KVAR)	APL (kW)	VD	VSI	LR (%)
proposed	MRFO	11 18 61	598.0106 425.9067 1895.7	196.5566 139.9888 623.1016	20.7702	0.00016	0.9772	90.77
[51]	QODELFA	11 18 61	579.7 434.0 1901.3	190.5 142.6 624.9	20.774	0.00015	0.9770	90.77
[52]	SFSA	11 17 61	5435 4132 1.8728	1786 1358 6156	20.727	0.000330	0.9772	90.79
[56]	MOHHO	23 60 62	519 1176 1179	171 387 387	30.2	0.001	0.98	86.5742
[56]	MOIHHO	13 61 63	1083 799 1229	341 263 404	28.9	0.0003	0.98	87.1521

5.2.3. Three DG Units at 0.82 Power Factor

In this case, the injected reactive power is higher compared to previous cases. The obtained results are improved significantly. The power loss characteristics shown in Figure 26 show an improved performance. The power loss at branch 56 is minimized to 0.0091 kW. The minimum voltage recorded at bus 65 is maximized to be 0.9976 as shown in Figure 27. Finally, the VSI simulated in Figure 28 demonstrates that the minimum VSI at branch 64 is maximized to 0.9905. This case provides better performance of the power system due to the increment of reactive power injected into the system.

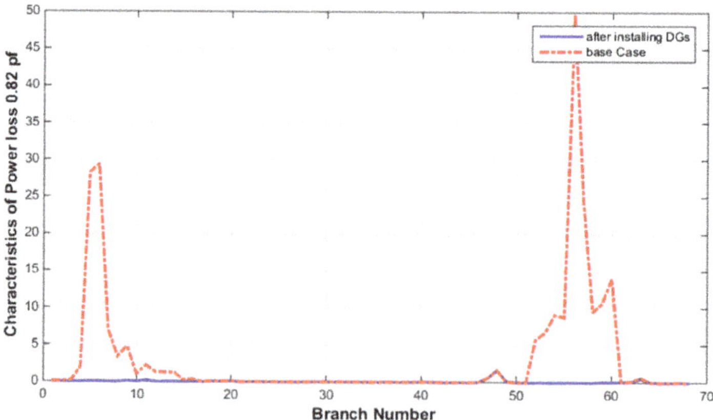

Figure 26. Characteristics of power loss at 0.82 pf.

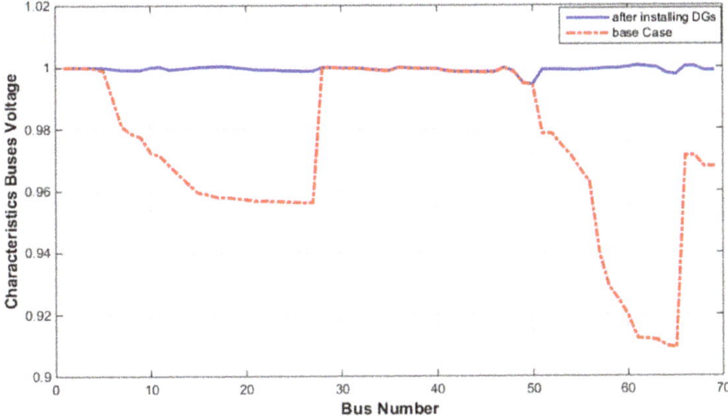

Figure 27. Characteristics of voltage at 0.82 pf.

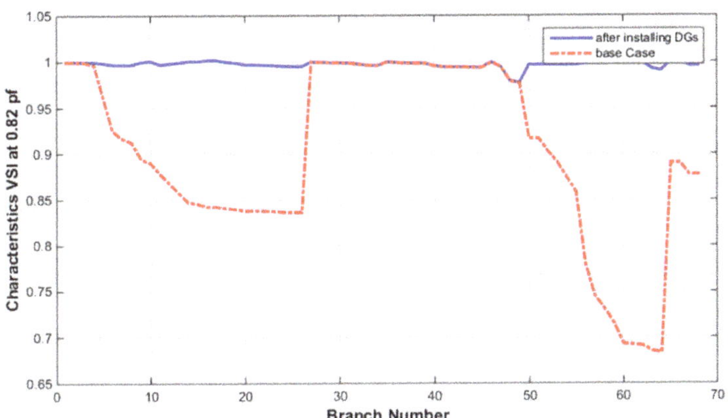

Figure 28. Characteristics of VSI at 0.82 pf.

Table 8 shows the results of optimizing DG units based on MRFO and QODELFA techniques. The given results show the effectiveness of the proposed algorithm and the other ones. The selected buses to install DG units are 11, 61, and 18. Power loss is minimized to be 4.2952, with a total loss reduction of 98.09%. The voltage deviation is minimized to its minimum value, namely 0.0001. Finally, the voltage stability index is enhanced to reach 0.9773 slightly higher than that for QODELFA. The MRFO algorithm provides almost the same results obtained by QODELFA with no superiority. QODELFA seems to be as good as MRFO, but it takes four times the number of iterations taken by MRFO. Increasing the number of iterations means increasing the total time of simulations, which decreases the effectiveness of the algorithm.

Table 8. 69-bus, three DG units, 0.82 pf.

Reference	Method	Location	Size (kW)	Size (KVAR)	APL (kW)	VD	VSI	LR (%)
proposed	MRFO	11 61 18	505.5087 1692.7 382.3651	352.8473 1181.5 266.8925	4.2952	0.00010	0.9773	98.09
[51]	QODELFA	11 18 61	505.8 385.9 1693.9	353.1 258.9 1182.3	4.297	0.00010	0.9771	98.09

5.2.4. Three DG Units at Optimum Power Factor

The main power system characteristics, power loss, voltage magnitude, and voltage stability index optimized by the MRFO algorithm are compared to the base case. The obtained results are demonstrating that MRFO is a powerful algorithm. The power loss for each branch is shown in Figure 29. The maximum power loss recorded at branch 56 is maximized to be 0.0056 kW. The new maximum power loss recorded at branch 48 with value 1.6312 kW. The minimum voltage magnitude recorded at bus 65 is enhanced to 0.9975 as shown in Figure 30. The minimum VSI recorded at branch 64 is maximized to 0.99. The new minimum VSI recorded at branch 49 has a value of 0.9773 as shown in Figure 31. This case gives the best solution compared to the former cases.

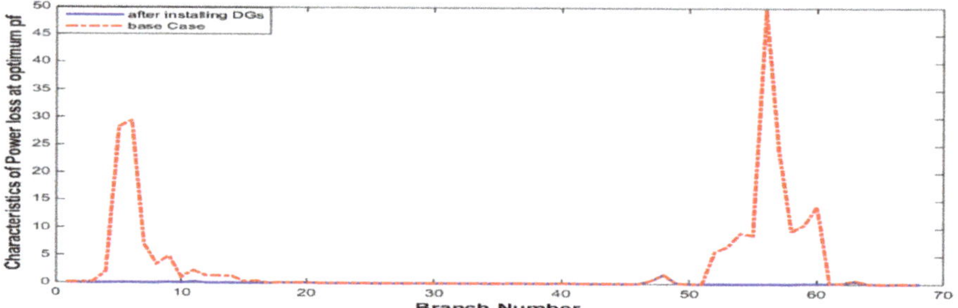

Figure 29. Characteristics at power loss at optimum pf.

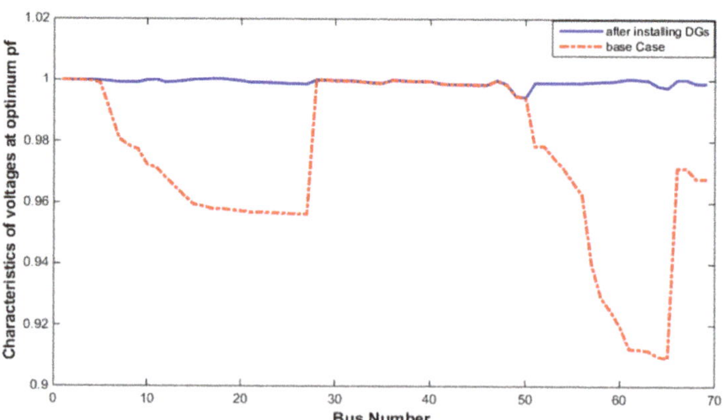

Figure 30. Characteristics of voltage at optimum pf.

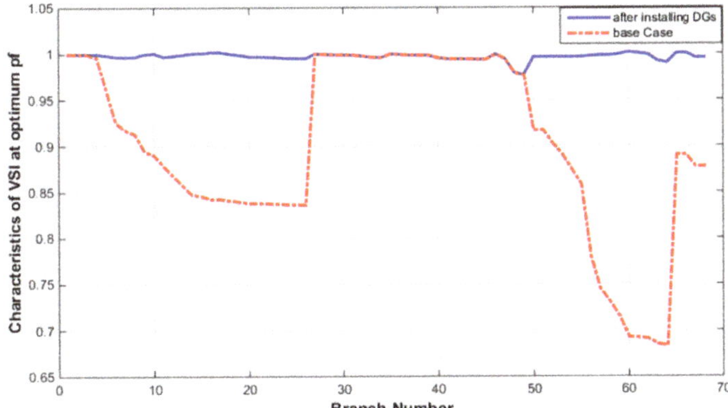

Figure 31. Characteristics of VSI at optimum pf.

The final sub-case in the IEEE 69-bus system is optimizing the size and location of three DG units operating at an optimum power factor. The results obtained by MRFO and three other techniques are listed in Table 9. The best locations determined using MRFO are 17, 11, and 61 which are completely different from the results of other techniques. The maximum power loss is minimized to 4.2775 kW with a total reduction of 98.0984% which is the best-obtained value. The voltage deviation also provided by the MRFO algorithm is the minimum value of 0.0001. Finally, the voltage stability index obtained by the MRFO algorithm is almost the same as the results obtained by SFSA. MRFO and SFSA techniques provide the best results compared to others. The MRFO algorithm takes a smaller number of iterations, namely 50, to achieve the optimum solution, unlike SFSA that takes double the number of iterations, 100. That gives superiority to the MRFO algorithm.

Table 9. 69-bus, three DG units at optimum pf.

Reference	Method	Location	DG Capacity and PF			APL (kW)	VD	VSI	LR (%)
			(kW)	(KVAR)	(pf)				
Proposed	MRFO	17 11 61	388.009 494.701 1680.9	254.684 365.035 1203.00	0.8360 0.8047 0.8132	4.2775	0.000106	0.9773	98.0984
[52]	SFSA	11 21 61	566.9 336.0 1675.2	397.0 222.7 1178.8	0.819 0.833 0.818	4.298	0.000116	0.9773	98.0892
[56]	MOHHO	15 60 61	332 314 1784	846 838 335	0.37 0.35 0.98	21.8	0.0008	0.98	90.3085
[56]	MOIHHO	13 49 62	1064 1235 1610	779 403 1181	0.81 0.95 0.81	13.9	0.0005	0.991	93.8206

The former four cases are compared to each other to investigate the effect of varying power factor on the performance of the system. Table 10 provides results obtained by the MRFO algorithm and represents the operating DG units at different power factors. The overall performance of the system is enhanced when moving from the unity power factor towards the optimum values. When all the DG units operate at an optimum power factor, the power loss was at its minimum of 4.2775 kW. In addition, the voltage deviation was at its minimum value of 0.0001. Finally, the voltage stability index was maximized to its higher value of 0.9773, the same as for 0.82 power factor operating conditions.

Table 10. Comparing results for different power factors.

Method	Location	DG Capacity and PF			APL (kW)	VD	VSI	LR (%)
		(kW)	(KVAR)	(Pf)				
Optimum pf	17 11 61	388.009 494.701 1680.9	254.684 365.035 1203.00	0.8360 0.8047 0.8132	4.2775	0.000106	0.9773	98.0984
0.82 pf	11 61 18	505.5087 1692.7 382.3651	352.8473 1181.5 266.8925	0.82 0.82 0.82	4.2952	0.00010	0.9773	98.09
0.95 pf	11 18 61	598.0106 425.9067 1895.7	196.5566 139.9888 623.1016	0.95 0.95 0.95	20.7702	0.00016	0.9772	90.77
Unity pf	18 11 61	530.0261 619.4599 1920.7	0 0 0	1 1 1	72.7496	0.0016	0.9522	67.6582

6. Conclusions and Future Work

A new meta-heuristic algorithm inspired by nature, namely the manta ray foraging optimization algorithm (MRFO) is applied to optimize the size and location of three DG units. Two standard power systems, 33-bus and 69-bus, are optimized based on three objective functions: power loss, voltage deviation, and voltage stability index. The proposed MRFO was tested successfully on the IEEE 33-bus system, and the IEEE 69-bus system to improve the voltage stability index, reduce voltage deviation, and decrease power losses. To investigate the performance of the proposed technique, four cases are applied to each power system categorized based on the operating power factor of the DG units. The MRFO algorithm has been compared to other techniques for the same operating system and the same operating conditions. It can be concluded that the MRFO algorithm is an effective and powerful technique. Compared to other techniques, such as: QODELFA, SFSA, CTLPO, CTLPO (ε constraint), MOHHO, MOIHHO, MOPSO, and MOWOA, it provides superiority in two cases: unity and 0.95 lagging power factor. In addition, the MRFO algorithm provides almost the same results for lagging power factor and optimum power factor for the 69 bus system. Like all other techniques, the MRFO algorithm could not provide superiority in all cases; SFSA provides better results in optimum power factor for the 33 bus system. Taking into consideration the number of iterations for all optimization algorithms, superiority would be assigned to MRFO. MRFO takes only 50 iterations to achieve an optimum value which is half and one-fourth the number of iterations taken by SFSA and QODELFA respectively. Operating DG units at unity power factor gives the worst results of all cases due to the lack of injected reactive power. Decreasing power factor means increasing injected reactive power which improves the system performance. But decreasing power factor must not exceed the optimum values of power factor to avoid over-voltage. When the DG units operate at an optimum power factor, they give the best results of all. Although the weight sum is an effective method to deal with multi-objective problems, it must be constrained to bigger objective functions such as cost function, in order to be easily compared with different references. The MRFO algorithm is a powerful technique that must be evaluated for much more complicated power systems especially a real type one.

Author Contributions: Conceptualization, M.G.H. and S.A.; methodology, M.G.H.; software, M.G.H.; validation, A.-A.A.M., A.A.I. and T.S.; formal analysis, M.G.H.; investigation, M.G.H. and S.A.; resources, M.G.H. and S.A.; data curation, A.-A.A.M. and A.A.I.; writing—original draft preparation, M.G.H.; writing—review and editing, M.G.H., S.A.; supervision, A.-A.A.M and A.A.I.; All authors have read and agreed to the published version of the manuscript.

Funding: This research received no external funding.

Conflicts of Interest: The authors declare no conflict of interest.

References

1. Twaha, S.; Ramli, M.A.M. A review of optimization approaches for hybrid distributed energy generation systems: Off-grid and grid-connected systems. *Sustain. Cities Soc.* **2018**, *41*, 320–331. [CrossRef]
2. Mehigan, L.; Deane, J.; Gallachóir, B.; Bertsch, V. A review of the role of distributed generation (DG) in future electricity systems. *Energy* **2018**, *163*, 822–836. [CrossRef]
3. Singh, A.; Parida, S. A review on distributed generation allocation and planning in deregulated electricity market. *Renew. Sustain. Energy Rev.* **2018**, *82*, 4132–4141. [CrossRef]
4. Muruganantham, B.; Gnanadass, R.; Padhy, N.P. Challenges with renewable energy sources and storage in practical distribution systems. *Renew. Sustain. Energy Rev.* **2017**, *73*, 125–134. [CrossRef]
5. Evangelopoulos, V.A.; Georgilakis, P.S.; Hatziargyriou, N.D. Optimal operation of smart distribution networks: A review of models, methods and future research. *Electr. Power Syst. Res.* **2016**, *140*, 95–106. [CrossRef]
6. Sampangi, K.S. A review on optimal allocation and sizing techniques for DG in distribution systems. *Int. J. Renew. Energy Res. (IJRER)* **2018**, *8*, 1236–1256.
7. Georgilakis, P.S.; Hatziargyriou, N.D. Optimal distributed generation placement in power distribution networks: Models, methods, and future research. *IEEE Trans. Power Syst.* **2013**, *28*, 3420–3428. [CrossRef]
8. Ansari, M.M.; Guo, C.; Shaikh, M.S.; Jatoi, M.A.; Yang, C.; Zhang, J. A Review of Technical Methods for Distributed Systems with Distributed Generation (DG). In Proceedings of the 2019 2nd International Conference on Computing, Mathematics and Engineering Technologies (iCoMET), Sukkur, Sindh, Pakistan, 30–31 January 2019; pp. 1–7.
9. Acharya, P.S. Intelligent algorithmic multi-objective optimization for renewable energy system generation and integration problems: A review. *Int. J. Renew. Energy Res. (IJRER)* **2019**, *9*, 271–280.
10. Alkhalaf, S.; Senjyu, T.; Saleh, A.A.; Hemeida, A.M.; Mohamed, A.-A.A. A MODA and MODE comparison for optimal allocation of distributed generations with different load levels. *Sustainability* **2019**, *11*, 5323. [CrossRef]
11. Alsadi, S.; Khatib, T. Photovoltaic power systems optimization research status: A review of criteria, constrains, models, techniques, and software tools. *Appl. Sci.* **2018**, *8*, 1761. [CrossRef]
12. Ming, M.; Wang, R.; Zha, Y.; Zhang, T. Multi-objective optimization of hybrid renewable energy system using an enhanced multi-objective evolutionary algorithm. *Energies* **2017**, *10*, 674.
13. Khaled, U.; Eltamaly, A.M.; Beroual, A. Optimal power flow using particle swarm optimization of renewable hybrid distributed generation. *Energies* **2017**, *10*, 1013. [CrossRef]
14. Mohamed, A.-A.; Ali, S.; Alkhalaf, S.; Senjyu, T.; Hemeida, A.M. Optimal Allocation of Hybrid Renewable Energy System by Multi-Objective Water Cycle Algorithm. *Sustainability* **2019**, *11*, 6550. [CrossRef]
15. Yang, B.; Guo, Y.; Xiao, X.; Tian, P. Bi-level Capacity Planning of Wind-PV-Battery Hybrid Generation System Considering Return on Investment. *Energies* **2020**, *13*, 3046. [CrossRef]
16. Kumar, M.; Nallagownden, P.; Elamvazuthi, I. Optimal placement and sizing of renewable distributed generations and capacitor banks into radial distribution systems. *Energies* **2017**, *10*, 811. [CrossRef]
17. Hemeida, A.; El-Ahmar, M.; El-Sayed, A.; Hasanien, H.M.; Alkhalaf, S.; Esmail, M.; Senjyu, T. Optimum design of hybrid wind/PV energy system for remote area. *Ain Shams Eng. J.* **2020**, *11*, 11–23. [CrossRef]
18. Susowake, Y.; Masrur, H.; Yabiku, T.; Senjyu, T.; Howlader, A.M.; Abdel-Akher, M.; Hemeida, A.M. A multi-objective optimization approach towards a proposed smart apartment with demand-response in Japan. *Energies* **2019**, *13*, 127. [CrossRef]
19. Gamil, M.M.; Sugimura, M.; Nakadomari, A.; Senjyu, T.; Howlader, H.O.R.; Takahashi, H.; Hemeida, A.M. Optimal Sizing of a Real Remote Japanese Microgrid with Sea Water Electrolysis Plant Under Time-Based Demand Response Programs. *Energies* **2020**, *13*, 3666. [CrossRef]
20. Howlader, H.; Adewuyi, O.B.; Hong, Y.-Y.; Mandal, P.; Hemeida, A.M.; Senjyu, T. Energy Storage System Analysis Review for Optimal Unit Commitment. *Energies* **2019**, *13*, 158. [CrossRef]
21. Saleh, A.A.; Mohamed, A.-A.A.; Hemeida, A.M. Optimal Allocation of Distributed Generations and Capacitor Using Multi-Objective Different Optimization Techniques. In Proceedings of the 2019 International Conference on Innovative Trends in Computer Engineering (ITCE), Aswan, Egypt, 2–4 February 2019; pp. 377–383.

22. Mohamadi, M.R.; Abedini, M.; Rashidi, B. An adaptive multi-objective optimization method for optimum design of distribution networks. *Eng. Optim.* **2019**, *52*, 194–217. [CrossRef]
23. Aranizadeh, A.; Niazazari, I.; Mirmozaffari, M. A novel optimal distributed generation planning in distribution network using cuckoo optimization algorithm. *Eur. J. Electr. Eng. Comput. Sci.* **2019**, *3*, 1–5. [CrossRef]
24. Musa, A.; Maledo, E.D. Review of Metaheuristic Optimization Methods for Optimal Location and Sizing of Distributed Generation. *J. Eng. Res. Appl.* **2019**, *9*, 57–65.
25. Trivedi, G.; Markana, A.; Bhatt, P.; Patel, V. Optimal Sizing and Placement of Multiple Distributed Generators using Teaching Learning Based Optimization Algorithm in Radial Distributed Network. In Proceedings of the 2019 6th International Conference on Control, Decision and Information Technologies (CoDIT), Paris, France, 23–26 April 2019; pp. 958–963.
26. Saleh, A.A.; Mohamed, A.-A.A.; Hemeida, A.M.; Ibrahim, A.A. Comparison of Different Optimization Techniques for Optimal Allocation of Multiple Distribution Generation. In Proceedings of the 2018 International Conference on Innovative Trends in Computer Engineering (ITCE), Aswan, Egypt, 19–21 February 2018; pp. 317–323.
27. Jegadeesan, M.; Babulal, C.K.; Subathra, V. Simultaneous Placement of Multi-DG and Capacitor in Distribution System Using Hybrid Optimization Method. In Proceedings of the 2017 International Conference on Innovations in Green Energy and Healthcare Technologies (IGEHT), Coimbatore, India, 16–18 March 2017; pp. 1–7.
28. Ton, T.N.; Nguyen, T.T.; Truong, A.V.; Vu, T.P. Optimal Location and Size of Distributed Generators in an Elecric Distribution System based on a Novel Metaheuristic Algorithm. *Eng. Technol. Appl. Sci. Res.* **2020**, *10*, 5325–5329.
29. Ang, S.; Leeton, U. Optimal Placement and Size of Distributed Generation in Radial Distribution System Using Whale Optimization Algorithm. *Suranaree J. Sci. Technol.* **2019**, *26*, 1–12.
30. Maurya, H.; Bohra, V.K.; Pal, N.S.; Bhadoria, V.S. Effect of Inertia Weight of PSO on Optimal Placement of DG. In *IOP Conference Series: Materials Science and Engineering*; IOP Publishing: Bristol, UK, 2019; Volume 594, p. 012011. [CrossRef]
31. Kashyap, M.; Mittal, A.; Kansal, S. Optimal Placement of Distributed Generation Using Genetic Algorithm Approach. In *Proceeding of the Second International Conference on Microelectronics, Computing & Communication Systems (MCCS 2017)*; Springer: Singapore, 2019; pp. 587–597.
32. Nguyen, T.P.; Tran, T.T.; Vo, D.N. Improved stochastic fractal search algorithm with chaos for optimal determination of location, size, and quantity of distributed generators in distribution systems. *Neural Comput. Appl.* **2018**, *31*, 7707–7732. [CrossRef]
33. Mohamed, A.-A.A.; Ali, S.M.; Hemeida, A.; Ibrahim, A.A. Optimal Placement of Distributed Energy Resources Including Different Load Models Using Different Optimization Techniques. In Proceedings of the 2019 International Conference on Innovative Trends in Computer Engineering (ITCE), Aswan, Egypt, 2–4 February 2019; pp. 389–395.
34. Abdelaziz, M.; Moradzadeh, M. Monte-Carlo simulation based multi-objective optimum allocation of renewable distributed generation using OpenCL. *Electr. Power Syst. Res.* **2019**, *170*, 81–91. [CrossRef]
35. Arulraj, R.; Kumarappan, N. A novel multi-objective hybrid WIPSO-GSA algorithm-based optimal DG and capacitor planning for techno-economic benefits in radial distribution system. *Int. J. Energy Sect. Manag.* **2019**, *13*, 98–127.
36. Abass, Y.; Abido, M.; Al-Muhaini, M.; Khalid, M. Multi-Objective Optimal DG Sizing and Placement in Distribution Systems Using Particle Swarm Optimization. In Proceedings of the 2019 IEEE Innovative Smart Grid Technologies-Asia (ISGT Asia), Chengdu, China, 21–24 May 2019; pp. 1857–1861.
37. Agajie, T.F.; Salau, A.O.; Hailu, E.A.; Sood, M.; Jain, S. Optimal Sizing and Siting of Distributed Generators for Minimization of Power Losses and Voltage Deviation. In Proceedings of the 2019 5th International Conference on Signal Processing, Computing and Control (ISPCC), Solan, India, 10–12 October 2019; pp. 292–297.
38. Mustaffa, S.S.; Musirin, I.; Zamani, M.M.; Othman, M. Pareto optimal approach in Multi-Objective Chaotic Mutation Immune Evolutionary Programming (MOCMIEP) for optimal Distributed Generation Photovoltaic (DGPV) integration in power system. *Ain Shams Eng. J.* **2019**, *10*, 745–754. [CrossRef]
39. Liu, W.; Luo, F.; Liu, Y.; Ding, W. Optimal Siting and Sizing of Distributed Generation Based on Improved Nondominated Sorting Genetic Algorithm II. *Processes* **2019**, *7*, 955. [CrossRef]
40. SriKumar, K. Optimal Size & Location of Distributed Generation using Bird Swarm Optimization with Cuckoo Search Sorting Algorithm. *Int. J. Eng. Adv. Technol. (IJEAT)* **2019**, *8*, 781–787.

41. Ali, S.M.; Mohamed, A.-A.A.; Hemeida, A. A Pareto Strategy based on Multi-Objective for Optimal Placement of Distributed Generation Considering Voltage Stability. In Proceedings of the 2019 International Conference on Innovative Trends in Computer Engineering (ITCE), Aswan, Egypt, 2–4 February 2019; pp. 498–504.
42. Saad, M.A.; El-Ghany, H.A.A.; Azmy, A.M. Optimal DG deployment to improve voltage stability margin considering load variation. In Proceedings of the 2017 Nineteenth International Middle East Power Systems Conference (MEPCON), Nasr City, Egypt, 19–21 December 2017; pp. 765–771.
43. Bouketir, O.; Sebaa, H.; Bouktir, T. Optimal Integration of Distributed Generation in a Radial Distribution System Using BW/FW Sweep and PSO Algorithm. *Adv. Eng. Forum* **2020**, *35*, 94–101. [CrossRef]
44. Ali, S.M.; Mohamed, A.-A.A.; Hemeida, A. Impact of Optimal Allocation of Renewable Distributed Generation in Radial Distribution Systems Based on Different Optimization Algorithms. In Proceedings of the 2019 International Conference on Innovative Trends in Computer Engineering (ITCE), Aswan, Egypt, 2–4 February 2019; pp. 472–478.
45. Duong, M.Q.; Pham, T.D.; Nguyen, T.T.; Doan, A.T.; Van Tran, H. Determination of optimal location and sizing of solar photovoltaic distribution generation units in radial distribution systems. *Energies* **2019**, *12*, 174. [CrossRef]
46. Abul'Wafa, A.R. Ant-lion optimizer-based multi-objective optimal simultaneous allocation of distributed generations and synchronous condensers in distribution networks. *Int. Trans. Electr. Energy Syst.* **2018**, *29*, e2755. [CrossRef]
47. Nagaballi, S.; Kale, V.S. Optimal Deployment of DG and DSTATCOM in Distribution System Using Swarm Intelligent Techniques. In Proceedings of the 2018 International Conference on Robotics, Control and Automation Engineering, Beijing, China, 26–28 December 2018; pp. 65–70.
48. Prakash, D.; Lakshminarayana, C. Multiple DG placements in radial distribution system for multi objectives using Whale Optimization Algorithm. *Alex. Eng. J.* **2018**, *57*, 2797–2806. [CrossRef]
49. Peesapati, R.; Yadav, V.K.; Kumar, N. Flower pollination algorithm based multi-objective congestion management considering optimal capacities of distributed generations. *Energy* **2018**, *147*, 980–994. [CrossRef]
50. Saleh, A.A.; Mohamed, A.-A.A.; Hemeida, A.M.; Ibrahim, A.A. Multi-Objective Whale Optimization Algorithm for Optimal Allocation of Distributed Generation and Capacitor Bank. In Proceedings of the 2019 International Conference on Innovative Trends in Computer Engineering (ITCE), Aswan, Egypt, 2–4 February 2019; pp. 459–465.
51. Mahfoud, R.J.; Sun, Y.; Alkayem, N.F.; Alhelou, H.H.; Siano, P.; Shafie-Khah, M. A novel combined evolutionary algorithm for optimal planning of distributed generators in radial distribution systems. *Appl. Sci.* **2019**, *9*, 3394. [CrossRef]
52. Nguyen, T.P.; Vo, D.N. A novel stochastic fractal search algorithm for optimal allocation of distributed generators in radial distribution systems. *Appl. Soft Comput.* **2018**, *70*, 773–796. [CrossRef]
53. Sukhwal, P.B.; Singh, P. Optimization of Distributed Generation using Genetics Algorithm and Improvement in Multiobjective Function. *Int. Res. J. Eng. Technol. (IRJET)* **2019**, *6*, 370–377.
54. Quadri, I.A.; Bhowmick, S.; Joshi, D. A comprehensive technique for optimal allocation of distributed energy resources in radial distribution systems. *Appl. Energy* **2018**, *211*, 1245–1260. [CrossRef]
55. Ali, E.; Elazim, S.M.A.; Abdelaziz, A.Y. Optimal allocation and sizing of renewable distributed generation using ant lion optimization algorithm. *Electr. Eng.* **2018**, *100*, 99–109. [CrossRef]
56. Selim, A.; Kamel, S.; Alghamdi, A.S.; Jurado, F. Optimal Placement of DGs in Distribution System Using an Improved Harris Hawks Optimizer Based on Single-and Multi-Objective Approaches. *IEEE Access* **2020**, *8*, 52815–52829. [CrossRef]
57. Kamel, S.; Selim, A.; Jurado, F.; Yu, J.; Xie, K.; Yu, C. Multi-Objective Whale Optimization Algorithm for Optimal Integration of Multiple DGs into Distribution Systems. In Proceedings of the 2019 IEEE Innovative Smart Grid Technologies-Asia (ISGT Asia), Chengdu, China, 21–24 May 2019; pp. 1312–1317.
58. Ali, E.; Elazim, S.A.; Abdelaziz, A.Y. Ant Lion Optimization Algorithm for optimal location and sizing of renewable distributed generations. *Renew. Energy* **2017**, *101*, 1311–1324. [CrossRef]
59. Game, P.S.; Vaze, D. Bio-inspired Optimization: Metaheuristic algorithms for optimization. *arXiv* **2020**, arXiv:2003.11637.
60. Zhao, W.; Zhang, Z.; Wang, L. Manta ray foraging optimization: An effective bio-inspired optimizer for engineering applications. *Eng. Appl. Artif. Intell.* **2020**, *87*, 103300. [CrossRef]
61. Stevens, G.M.W. Conservation and Population Ecology of Manta Rays in the Maldives. Ph.D. Thesis, University of York, York, UK, 2016.

62. Rajendran, A.; Narayanan, K. Optimal installation of different DG types in radial distribution system considering load growth. *Electr. Power Compon. Syst.* **2017**, *3*, 739–751. [CrossRef]
63. Sharma, S.K.; Palwalia, D.K.; Shrivastava, V. Distributed generation integration optimization using fuzzy logic controller. *AIMS Energy* **2019**, *7*, 337–348. [CrossRef]
64. Kazeminejad, M.; Ghaffarianfar, M.; Hajizadeh, A. Optimal Sizing and Location of Distributed Generation Units to Improve Voltage Stability and Reduce Power Loss in the Distribution System. *J. Appl. Dyn. Syst. Control* **2019**, *2*, 41–47.
65. Selim, A.; Kamel, S.; Jurado, F. Efficient optimization technique for multiple DG allocation in distribution networks. *Appl. Soft Comput.* **2020**, *86*, 105938. [CrossRef]
66. Tolba, M.; Rezk, H.; Tulsky, V.; Diab, A.A.Z.; Abdelaziz, A.Y.; Vanin, A. Impact of optimum allocation of renewable distributed generations on distribution networks based on different optimization algorithms. *Energies* **2018**, *11*, 245. [CrossRef]
67. Rupa, M.; Ganesh, S. Power flow analysis for radial distribution system using backward/forward sweep method. *Int. J. Electr. Comput. Electron. Commun. Eng.* **2014**, *8*, 1540–1544.
68. Upper, N.; Hemeida, A.M.; Ibrahim, A.A. Moth-Flame Algorithm and Loss Sensitivity Factor for Optimal Allocation of Shunt Capacitor Banks in Radial Distribution Systems. In Proceedings of the 2017 Nineteenth International Middle East Power Systems Conference (MEPCON), Nasr City, Egypt, 19–21 December 2017; pp. 851–856.
69. Selem, S.I.; Hasanien, H.M.; El-Fergany, A.A. Parameters extraction of PEMFC's model using manta rays foraging optimizer. *Int. J. Energy Res.* **2020**, *44*, 4629–4640. [CrossRef]
70. Ghosh, K.K.; Guha, R.; Bera, S.K.; Kumar, N.; Sarkar, R. S-Shaped versus V-Shaped Transfer Functions for Binary Manta Ray Foraging Optimization in Feature Selection Problem. *Artif. Intell. Mach. Learn.* **2020**. [CrossRef]

Article

Distributed Machine Learning on Dynamic Power System Data Features to Improve Resiliency for the Purpose of Self-Healing

Miftah Al Karim [1], Jonathan Currie [2] and Tek-Tjing Lie [3,*]

1 The Lines Company, Te Kuiti 3910, New Zealand; miftah.ak@gmail.com
2 Rocket Lab, Auckland 1060, New Zealand; jonathan.currie@hotmail.com
3 School of Engineering, Computer and Mathematical Sciences, Auckland University of Technology, Auckland 1010, New Zealand
* Correspondence: tek.lie@aut.ac.nz

Received: 7 June 2020; Accepted: 2 July 2020; Published: 6 July 2020

Abstract: Numerous online methods for post-fault restoration have been tested on different types of systems. Modern power systems are usually operated at design limits and therefore more prone to post-fault instability. However, traditional online methods often struggle to accurately identify events from time series data, as pattern-recognition in a stochastic post-fault dynamic scenario requires fast and accurate fault identification in order to safely restore the system. One of the most prominent methods of pattern-recognition is machine learning. However, machine learning alone is neither sufficient nor accurate enough for making decisions with time series data. This article analyses the application of feature selection to assist a machine learning algorithm to make better decisions in order to restore a multi-machine network which has become islanded due to faults. Within an islanded multi-machine system the number of attributes significantly increases, which makes application of machine learning algorithms even more erroneous. This article contributes by proposing a distributed offline-online architecture. The proposal explores the potential of introducing relevant features from a reduced time series data set, in order to accurately identify dynamic events occurring in different islands simultaneously. The identification of events helps the decision making process more accurate.

Keywords: self-healing grid; machine-learning; feature extraction; event detection

1. Introduction

Self-healing is a lucrative feature in the service restoration process of a power system which in turn improves system resiliency. Power system resiliency often refers to the capacity of a system to maintain stability after a high impact events with minimum interruptions [1]. Such resiliency is key to achieve the objectivity of developing a Self-healing structure [2]. In the recent studies, under the category of post-fault islanding scenarios, self-healing mechanism is getting significant attention [3]. The prominent trend in addressing self-healing is established through the local or distributed control. It is because distributed control is quite effective in faster decision making. However, applying local or distributed control is quite dependent on the dynamic characteristics of the network under analysis. On the other hand, dynamic characteristics are heavily influenced by the pre-contingent stochastic parameters, fault analysis, demand response (DR) and also the performance of the algorithms assessing power system security [2,4].

In previous studies, different methods and systems have been investigated for fault detection, isolation, and service restoration (FDIR). Some of these methods are Central Controller, Distribution Automation System (DAS), Automatic Controlled Switching (ACSs), Fault Passage Indication (FPI), Corrective Voltage Control (CVC), Emergency Demand Response Program (EDRP) and more [5].

Most of these approaches are data-intensive programs. On the top of that, in a modern grid, the integration of renewable energy makes it quite impossible to comprehend the security of a system, on an online basis for all possible scenarios. Thus self-healing function can be considered a prime candidate by the systems based on machine learning algorithms, such as Artificial Neural Network (ANN), Support Vector Machine (SVM), Random forest. If the underlying events during any critical contingency are properly identified [6], the identification can lead towards developing intelligent self-healing strategies. The process can be developed through expert systems, with a large scenario based dataset. However, machine learning systems are affected by processing time and memory. Which, in a fast restoration system is not desired, especially for those grids that require adaptive online decision making. Besides, for a large network, performing the dynamic security assessment (DSA) is an increasingly complex problem affected by numerous criteria [7,8]. Under these contexts, the traditional applications of machine learning algorithms are progressively being less effective.

The modern trend dictates that the machine learning algorithms should be trained offline basis and implemented online [9]. But multi-machine systems generate large scale data. For many online DSA programs, this approach is unattractive. Therefore, previous studies often recommend alternate solutions such as dimensionality reduction, feature selection [10–12]. Feature selection process is implemented to modify a data set to increase the accuracy of prediction [13,14]. However, it adds redundancy in the algorithmic steps [15]. In a large-scale power system, such redundancy will be affected by the curse of dimensionality [12]. Reference [14] mitigates this problem using an energy function based feature selection. The proposed method outperforms the method that uses raw inputs to train the machine learning algorithms, considering a smaller dataset for both the cases. This strategy proves to be highly effective, but a smaller dataset is insufficient while addressing multiple stochastic scenarios. Because in a hybrid grid, stochastic parameters make it challenging to assess dynamic security using analytical approaches. One solution to address this challenge is a Monte Carlo-based simulation method. This approach can establish the stability boundary of a grid with multiple stochastic parameters [16]. Once the stability boundary is established for a coherent group of stochastic parameters, a limited data set can be used for online DSA programs. Another approach to address the curse of dimensionality is to implement dimensionality reduction based event detection methods [10]. This method can successfully differentiate oscillatory and non-oscillatory scenarios. However, the loss of other valuable information, makes it difficult for such a simplistic approach alone to address the problem of event-detection in a hybrid system [9,17].

Due to such reasons, this work proposes a method of adding a layer of extracted features from the reduced data set. The underlying motivation is to bridge this gap between dimensionality reduction and loss of information. The power system data is collected from the available generators and transmission lines under different contingencies. The data is then reduced and features are extracted. The features are then used as attributes to train a machine learning algorithm. These features help the machine learning algorithm make accurate classification under those stochastic scenarios. The performance of the algorithm has been tested through an event-based decision making process to restore a segmented grid that significantly improves system reliability. The novelty of the process has been further justified by comparing the proposed algorithm with some existing methods. In summary, the overall contributions of this work are as follows:

1. Development of a feature selection based event detection algorithm for a multi-machine based grid, that can be sectionalized under duress. The proposed algorithm is prepared for the segmented power system used in this study. Despite not being a generic solution for all types of grids the algorithm introduces novelty in the decision making process.
2. In larger systems, the curse of dimensionality poses a bigger threat in applying machine learning algorithms, specially for making decisions. The proposed method, by implementing feature extraction on a reduced data set, addresses those challenges and enables an effective decision making scheme.

2. System Under Consideration

An IEEE-39 bus 10-machine test system as shown in Figure 1 is used in this study [18]. During a rotor angle instability the 39-bus test system can be divided into several independent islands. In this study these islands are considered as independent and locally controlled multi-machine systems when disconnected. In order to establish a relationship between the generator model and the constant impedance load model, classical energy functions were used [14];

$$M_i \frac{d^2\delta_i}{dt^2} + D_i \frac{d\delta_i}{dt} = P_i - \sum_{j=1, j \neq i}^{m} (C_{ij} \sin \delta_i j + D_{ij} \cos \delta_{ij}), \tag{1}$$

where P_i is the active power of the i-th machine, M_i is the moment of inertia of the i-th machine, D_i is the damping co-efficient, m is the number of synchronous generators. δ_i is the rotor angle of the i-th generator. C_{ij} and D_{ij} are the function of transfer conductance and susceptance of the reduced network. The per unit inertia constant of each of these generators is $H = \frac{1}{2}M\omega$, where ω is the synchronous speed.

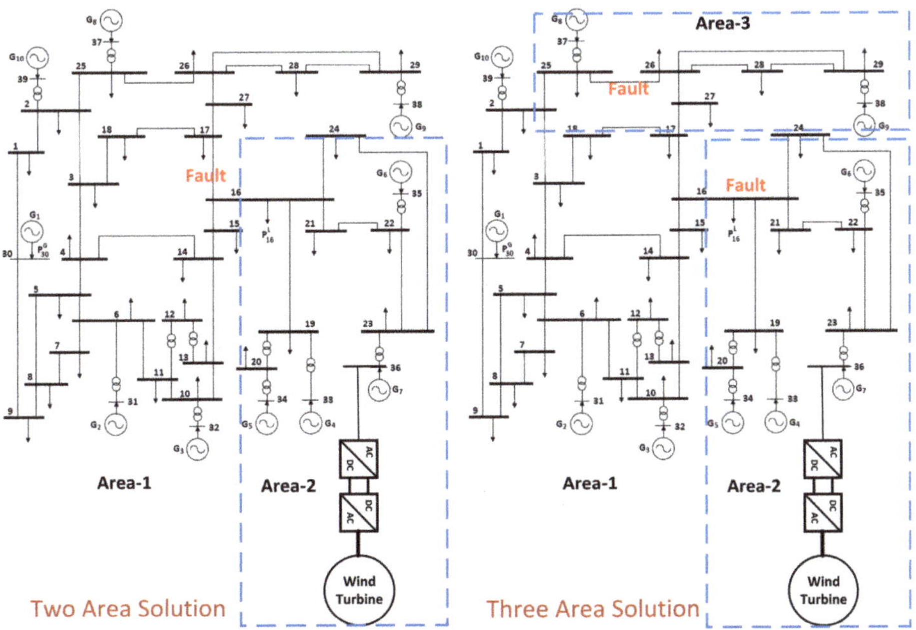

Figure 1. The sectionalized grid model. By disconnecting the transmission lines multiple areas can be created.

The proposed power system does not to have any energy storage. It considers a conventional droop characteristics for the primary frequency and voltage control. The primary control in a standalone grid that does not have any storage unit, may cause frequency deviation even if the system remains at a steady state condition. Therefore, a secondary control is also imposed that acts on the deviations observed over the primary control and tries to balance the system. The secondary control has a slower dynamics than the primary control, the idea behind that is to introduce a decoupling mechanism between these two [19].

In order to do data analysis the dynamic data is produced from the synchronous generators. The generators are controlled by a multi-band power system stabilizer (PSS) model and a governor [20,21]. The governor of each generator is modeled as a tandem-compound steam prime mover system. The overall simulation is carried out using Matlab-Simulink [22,23]. PSS are the local controllers on synchronous machines. These are quite effective in eliminating system oscillations. However, tuning the set-points for a PSS, is usually carried out via analytical calculations. In a hybrid system, where the distributed energy generation and consumption are rapidly varied, a critical three-phase fault can easily disrupt the process of damping the oscillation. It is because analytical calculations often do not capture the stochastic nature of standalone system. Therefore, a three tier hierarchical control, where the top layer deals with the probabilistic nature of a system, can bring significant improvement [19]. In this study, a supervised secondary control scheme is implemented that modifies the reference for active power generation in the turbine governor. This action helps reducing the rotor speed deviation, which is crucial for the pre-tuned set-points of the multiband-PSS, to yield result in a post-contingent scenario. The proposed supervised controller is based on machine learning algorithms. The supervised secondary controller has been implemented on each generator through a distributed architecture. This distribution is based on the machine data available in a post-contingent segment. The initial feedback data to trigger the supervised secondary control is taken from the deviation in rotor angle of the individual machine [24].

The system is designed to have rotor angle instability during a critical contingency. When such a contingency is noticed, the system can be divided into multiple operation-autonomous islands, which is necessary to eliminate the instability [25]. The method applies an unsupervised machine learning based controlled islanding mechanism that utilizes coherency based grouping from historical database. The method is further discussed in the later sections of this study.

The synchronous generators in each area implement real power frequency and reactive power voltage droop control. To represent non-dispatchable energy generation, a wind power plant based on an induction generator is considered and connected in bus-36, which is closer to the Generator-7 [26]. The system considers two types of loads, critical invariant and non-critical variable load which can be shed if required. The wind power generation and the non critical loads are considered as the key stochastic parameters. The variable load is lumped in buses-8, 24, 32 [27]. In order to create a system wide rotor angle instability, a critical short circuit fault have been introduced close to the bus-16 and bus-17 [28]. It makes the generators, swing against each other in groups. The event is captured in Figure 2. The top half of Figure 2 shows the rotor angle in one of the generators, during the occurrence post fault instability. The bottom half of Figure 2 shows the generator speeds right after the short circuit fault.

The energy function dictates that if a critical disturbance occurs in the system, mechanical and electrical power go out of balance. Depending on the disturbance, the change in rotor angle may introduce a rotor angle instability. In turn, the rotor angle instability introduces large voltage fluctuation in the transmission lines. The transmission line voltage fluctuations can also be spotted from the terminal voltage of the generators affected by this instability. In this study, the terminal voltage data has been used to develop an optimized active power corrective control (CC) to eliminate the rotor angle instability in each of the islands. The corrective control parameters or the active power from different generators varies with the available wind power and load [29–32].

The machine learning algorithm developed is based on different power system events associated to FDIR. Within the time-line of different events, by analyzing the dynamic data collected from the generators, a distributed supervised secondary control scheme is developed; to achieve $\frac{dE_G}{dt} \rightarrow 0$ (E_G = Generator terminal voltage); after a major disturbance. The supervised secondary machine control is a process that includes the above mentioned CC technique. The proposed method is a modified and extended approach carried out in Reference [33]. The modification is brought by introducing a feature extraction method. The system is designed to have both normal operating mode and self healing mode, in this study only the self healing mode is considered. The self healing mode is invoked once a rotor angle instability is observed, right after a critical fault is cleared. After a large number of offline simulations, for this model it is observed that, the rate of change in between $-5^{\circ}/$ to $+5^{\circ}/\mathrm{s}$ in rotor angle deviation indicates a critical rotor angle instability where the grid has to be operated in the self-healing mode. The observation is shown in Figure 3. After *1-second or 50 samples* of the fault, the breach in threshold is clearly visible in the figure.

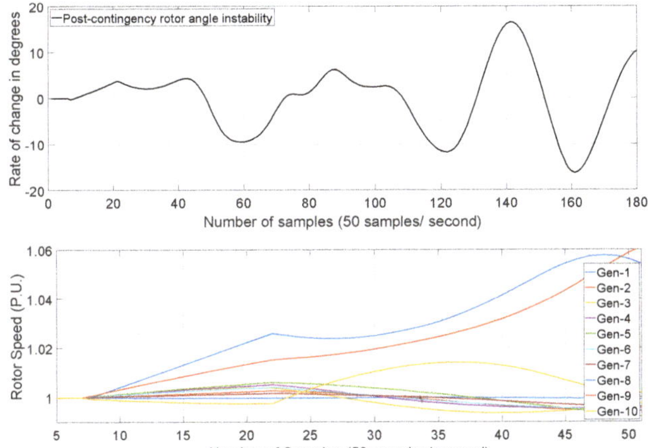

Figure 2. Top figure: Rotor angle instability observed once the critical fault is cleared. Bottom figure: Rotor speed fluctuation observed right after the three phase fault

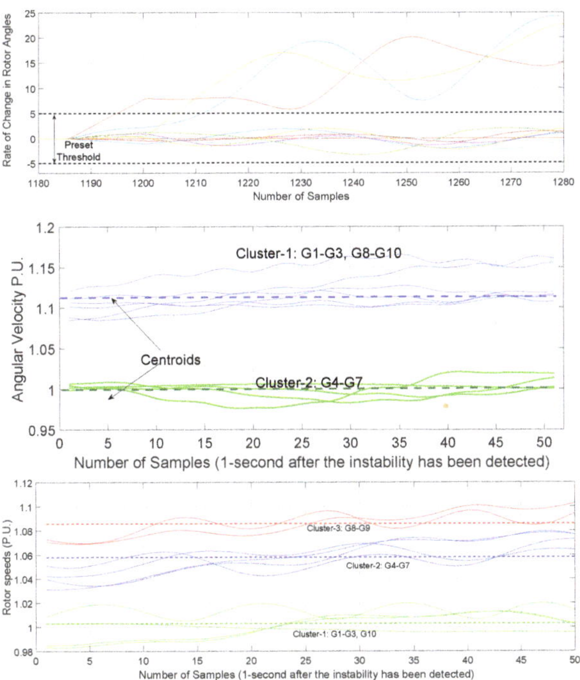

Figure 3. Top figure: High fluctuation in rotor angle due to the critical fault. Mid figure: K-means cluster of two groups of generators. Bottom figure: K-means cluster of three groups of generators.

The model in Figure 1 is used to generate time series dynamic data for training and testing the machine learning platform for the proposed supervised secondary controller. The proposed model is a modified IEEE-39 bus system, which is suitable for stability analysis. The demand and wind speed are randomly varied in order to do a Monte Carlo based simulation to prepare a stability database with the synchronous generator data. The dynamic parameters chosen are in per unit quantity and these are rotor speed ω, rotor angle deviation $d\delta$, active power generated P_E, and terminal voltage E_G. For simplicity, only the cases where the system can be stabilized and restored by dividing it into two segments are considered. During any contingencies these parameters from the generator, affected nodes is collected. These parameters are timeseries in nature. During the contingency from their dynamic behaviors, patterns in form of features are extracted. For example, the magnitude of a generator terminal voltage is considered dynamic during the contingency. Feature extraction is applied on that time series data in order to develop training and testing data for the next phases. A matrix of $444,037 \times 36$ is developed from the nine synchronous generators. Generator-2 is considered as the reference bus for calculating $d\delta$, therefore, the supervised control is not applied on this generator.

3. The Proposed Supervised Control

3.1. Controlled Islanding

Controlled islanding mechanism can be an effective way to mitigate system-wide instability and help restoration [25]. The method proposed in this study, applies an unsupervised machine learning algorithm on rotor-speed data in order to develop a coherency based grouping [25,34]. Power system can be modeled as a set of coupled differential algebraic equations as functions of rotor speed. Due to the nature of such coupling, rotor speed can often be used as an indicator for detecting coherent groups. An unsupervised clustering can be used as a tool to detect such groups [17]. At first, the process randomly selects centers for each cluster. The number of clusters is pre-defined. The membership of each data point is decided based on the objective function, that minimizes the Euclidean distance of each data point y_{ik} with the corresponding centroid z_i. Through this objective function the sum of all the Euclidean distances between every data point to their cluster-centroids.

$$D(y_{ik}.z_i) = \sum_{k=1}^{d} \sum_{i=1}^{C} \sqrt{(y_k - z_i)^2} \tag{2}$$

Here, d is the number of data points, $C = 1, 2, 3$ depending on the number of islands or no islanding required, z_i is the centroid of i-th cluster and y_k is the k-th data point. Once the D is calculated the centroids are re-selected and the objective function is repeated.

$$C_i = \frac{1}{n_i} \sum_{\forall y_p \in s_i} y_p \tag{3}$$

Here C_i is the center of the i-th cluster, n_i is the number of data points which in this case is the rotor-speed observed within an *1-second* data window after an instability is detected. y_p is the data vector after $p - th$ iteration. *1-second* data window is observed to be sufficient, to detect the coherency. The group is prepared, once the three phase fault is cleared but no restoration process or islanding has been implemented yet. In the Figure 3, the clusters are shown in three colors with their centroids. In middle section of the figure it is shown that, based on the coherency observed, two sets of generators have been prepared. *cluster-1* consists of {G1-G3, & G8-G10}, represents Area-1 and *cluster-2* consists of {G4-G7}, represents Area-2. The areas are created by disconnecting the transmission lines between node-(16&17) and node-(14&15). Similar clustering is also shown in the bottom part of the figure, with three groups of generators. The number of clusters are selected based on the available empirical data. During the training phase the clusters are prepared based on the coherency observed.

3.2. The Corrective Control

The primary control in an isolated grid may not be sufficient as it may cause frequency deviation after a non critical fault. Therefore, the secondary control is required, which implements a slower dynamics to fine tune the control parameters and prevents the frequency fluctuations. However, under a critical fault when the system wide rotor angle instability is observed, a secondary control often fails to stabilize the system [30,31]. This study implements a machine learning driven supervised secondary control during the post fault scenarios, where secondary control cannot maintain stability after a critical fault. The supervised secondary control is considered as the topmost part of a hierarchical control scheme [19,33,35].

The conventional droop characteristics satisfies the condition $D_{Pi}P_{Ni} = \Delta\omega_{max}$ and $D_{Qi}Q_{Ni} = \Delta E_{max}$ [3]. Where, i is the i-th stochastic scenario, N is the number of synchronous generators, P and Q are the generated active and reactive power, $\Delta\omega_{max}$ and ΔE_{max} are angular frequency and voltage deviations allowed. This study only considers scenarios where the allowed $\Delta\omega_{max}$ is exceeded. The dynamic response can be further understood through a linearization of the active and reactive power equations through a set of small signal models.

$$\Delta P(s) = \frac{G}{s + D_P G}\Delta\omega^*(s) \tag{4}$$

$$\Delta Q(s) = \frac{H}{1 + D_Q H}\Delta E^*(s) \tag{5}$$

Here, the coefficients, G and H depend on the nominal terminal voltage and the rotor angle of the generators and the transmission line voltages where power is transmitted. The secondary control scheme is evoked if the threshold of the $\Delta d\theta_{max}$ is crossed after a short circuit fault, as shown in Figure 3. At steady state conditions, under different stochastic scenarios, the voltage magnitudes and the angular differences between one generator and the point of common coupling is kept within a boundary by the PSS and governors. It results in a list of ΔP and ΔQ. Therefore, a Monte Carlo based simulation strategy is developed to prepare a stochastic database of the aforementioned P_{Ni}, Q_{Ni}, ΔP and ΔQ. During a rotor angle instability the supervised controller changes the active power references and keeps the active power generation fixed through the turbine governor. Due to this action the $\Delta\omega$ decreases and ΔP falls within the boundary suitable for the pre-tuned PSS. The database for optimized active power reference for different stochastic scenarios, is developed using genetic algorithms [2]. Overall this process is considered as the Corrective Control (CC). The CC is based on the following objective function:

$$Objective : min(\epsilon_{Vfault} = \sum_{j=1}^{M}[V_{prefault} - V_{postfault}(i)]^2). \tag{6}$$

If the steady state $\epsilon_{Vfault} \leq \epsilon_{threshold}$, the supervised control scheme is stopped. Here, j is the current data sample and M is the total number of data samples in that segment. For this study $M = 300$ equivalent to a five-second data window has been chosen. The objective function is subject to the constraint $0 \leq P_N \leq P_{N,max}$, $0 \leq P_{ls}$ where, N refers to the number of synchronous generators. Pls refers to the minimum temporary load shed. The overall system load is divided into two parts, variable demand and fixed demand. The temporary load shedding is carried out from the variable demand to maintain the energy balance; $\sum_{i=1}^{n_g} pg_i + \sum_{i=1}^{n_{wind}} pwind_i = P_{dpfl} + P_{Tloss} + P_{ls}$. Where, P_{dpfcl} is post fault load, P_{Tloss} is the transmission line loss, $pwind_i$ power generated from the wind power plant, n_g is number of synchronous generators, $n_{wind} = 1$. Once the system becomes stable the load is also restored. In this study, island area-1 requires temporary load shed when the wind power is relatively high. The overall workflow for the CC based control is shown in Figure 4. The workflow is divided into two phases, offline and online. The offline phase is based on a test and trial approach. The process initially starts with different stochastic scenarios. The CC is then applied to stabilize each segment and restore the grid. The CC is an optimization technique and the dynamic data obtained from the generators after CC has been applied is used for feature selection. These features represent different events and also the optimized decision parameters under that stochastic scenario. Once the machine learning algorithm is trained, it is then used in the online phase for making decisions for the CC. Being stochastic in nature, some of the predictions yields misclassification errors. Those data are used for further training.

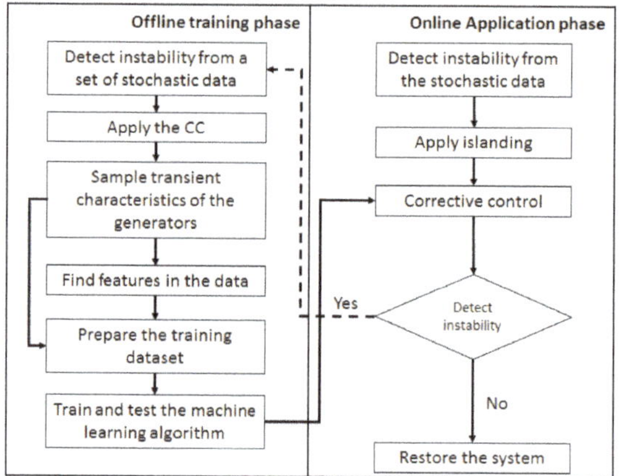

Figure 4. Workflow of the proposed method. This method is applied in a distributed architecture on each of the synchronous generators

Figure 5 shows the proposed model of the secondary control system. The supervised control is based on machine learning algorithms trained under different stochastic scenarios on the feature data to perform the proposed corrective control (CC) followed by islanding. The features are introduced in the consequent sections.

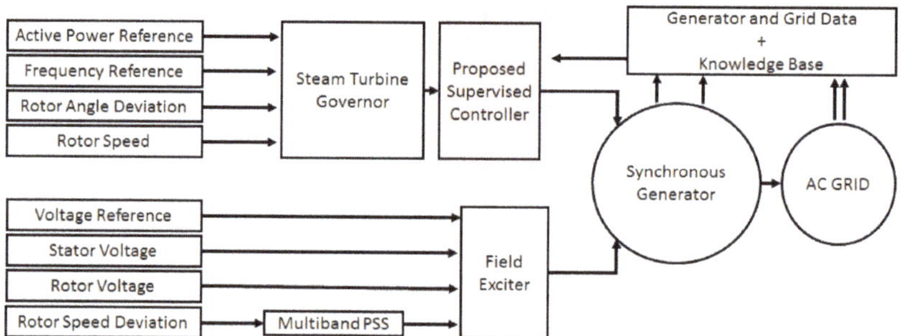

Figure 5. The governor and exciter control with an added functional block for the proposed secondary control.

3.3. Power System Events

Based on different wind power, loads, fault locations and number of faults; several cases have been observed. In each of these cases one or multiple critical three phase faults and consequently system wide rotor angle instability have been introduced. Based on these criteria the system has been either divided into two, three islands or no island at all. Table 1 shows a list of probable contingencies leading to three different islanding schemes.

Table 1. List of stochastic contingencies.

Decisions	Wind Speed (m/s)	Variable Demand (MW)	Contingencies
1	10	546	Loss of Genearator-4. No segmentation is required
2	12	609	Loss of Genearator-9. No segmentation is required
3	8	405	Three phase fault at Bus-16. Two area segmentation
4	12	35	Three phase fault at Bus-19. Two area segmentation
5	8.2	662	Three phase fault at Bus-16 and Bus-25. Three area segmentation
...	...	...	...
...	...	...	...
n	11.2	408	Three phase fault at Bus-25 and loss of Generator-4. Three area segmentation

Each of these cases has further been divided into multiple events namely, *Fault*, *Post fault rotor angle instability*, *Supervised control* and *Post restoration*. If the proposed system, fails to restore the grid, then the scenario-data is fed back to the offline training stage.

From the four above scenarios three time-lines are chosen, time-line for detecting post fault rotor angle instability, time-line for CC to stabilize the islanded network, and time-line representing the post restoration period. The two latter periods can either be stable or unstable. In Figure 6, all the different time-lines have been shown. Time-line-1 data is the candidate for decision making, time-line-2 is for applying CC and time-line-3 is the candidate data for evaluating the performance of the algorithm. Here, for a better understanding, terminal voltage of generator-7 has been chosen to demonstrate the events.

Furthermore, Figure 7 shows an example of the terminal voltage at Generator-7 under different scenarios (wind power and load), during the rotor angle instability. The similarity in the time series data is overwhelming, and that makes the application of a machine learning algorithm quite difficult [36]. However, each scenario has its own events and therefore, features can be used for distinguishing those events in each scenario.

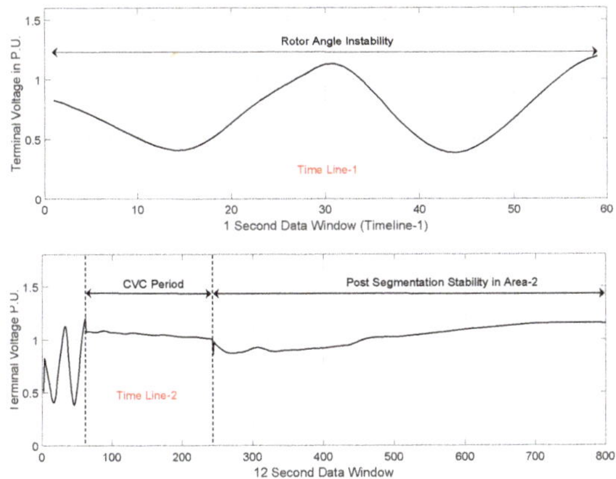

Figure 6. *Cont.*

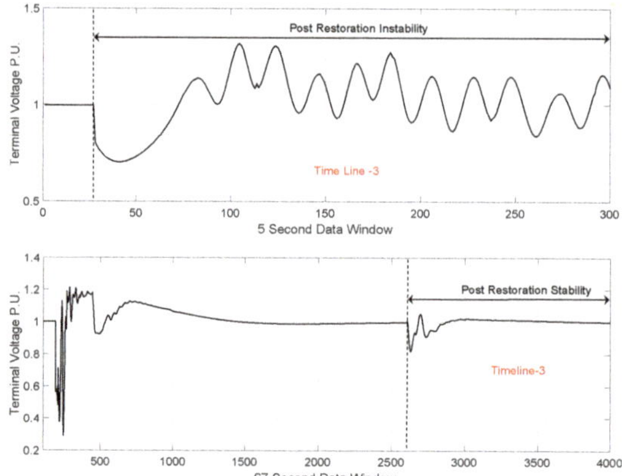

Figure 6. Top three figures are showing the three different time-lines. The bottom figure represents time-line-2&3 in the case when the proposed algorithm successfully restores the system.

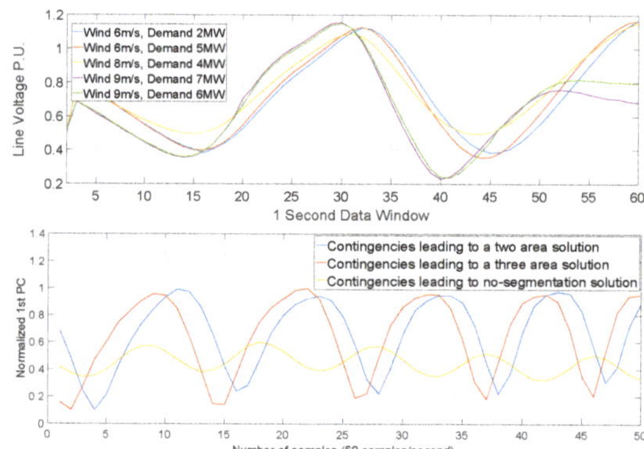

Figure 7. Top figure: Time series data of Generator-7 under different scenarios. Bottom figure: Normalized 1st principle component of 'timeline-1'under three different contingencies.

Like Figure 7, a similar variation in the phasor-signals is also observed in rotor speed and rotor angle from each synchronous generator. Such similarities in data makes it difficult to select a candidate for feature selection. Therefore, to retain information that can be crucial for the feature selection process, principle component analysis is carried out on these three types of phasor time series data [6]. The Principle Component Analysis (PCA) finds out the first component accounted for the most of the variations in generator data matrix. The variability is around 75%. The data matrix X has m number of observations with $n = 3$ number of attributes. The m-number of orthonormal basis functions are represented by W which is of $n \prod n$ dimensions. The data matrix is then represented as $X = TW'$, where T is $(t_{1,i} .....t_{m,i})$, which is used to form the coherent clusters for identifying principle components. Figure 8 shows that a significant variation in signal magnitudes and widths can be spotted in the reduced data (1st principle component) among different events. It further justifies that, the use of PCA for reducing the generator data and using that data for extracting features, is meaningful.

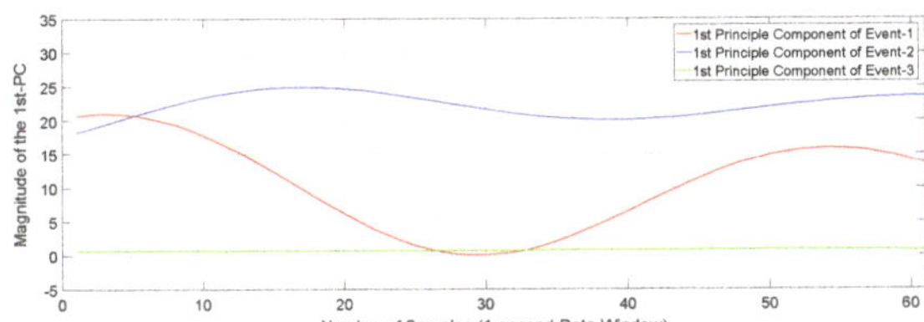

Figure 8. Variation Observed in the normalized 1st-PCs obtained from different events in generator-1 data. Each PC shown here, represents the near-end stage of one power system event.

3.4. Available Observed Features

As shown in Figure 8, the principle components have unique distinction in their shapes under different events. However, due to the nature of time series data, overlapping magnitudes confuses the learner. One solution is to use a large dataset for training. On the other hand this problem can significantly be overcome with the use of features [14,37]. Base on the discussion presented in earlier section, in this study three features, distinguishing each event, have been chosen. The selection process is mostly inspired by the observed variations in magnitude. Another motivation is drawn from the fact that, the inherent properties of a power system can provide significant information on a set of dependent variables based on a set of argument variables. This understanding is quite effective in detecting online phenomenon such as sensitivity. For example, active power can be used as a dependent-variable and node-voltage can be used as argument-variable to understand the voltage stability of a system [1]. Therefore, sensitivity data holds useful information. Moreover, feature extraction is a computationally expensive mechanism [37]. Thus, a set of simple and predefined features based on domain knowledge can reduce the computation burden. Keeping these issues in mind, this study selects the following three features:

1. Prominence of local maxima in terminal voltage [38–40]
2. Available frequencies in the time series voltage data [41]
3. Sensitivity $\partial P_E/\partial V_f$. Where, V_f is the voltage at the fault bus and ∂P_E is the active power generation from the subject generator [16].

Event detection and decision-making based on time-series data is not a predictive, rather a prescriptive analysis. In order to associate a feature with the underlying events in the time series data, a frame of reference has to be used. In this study the frame of reference is developed by using a sliding window technique on the time series data and recording the features. To do so, a *1-second* transition-window is selected to extract features and then those features are converted to different 'factors.' By 'factors' this study refers to a quantity that can represent a window using only one number. This action is necessary to represent the sliding windows using row vectors. After calculation, those factors are stored in the data-table. As mentioned earlier, each row signifies one attribute window. This table is later used to train the machine learning algorithm.

To prepare the features in each data window, sixty samples per second have been considered. The variation in magnitude of the first principle component has been considered as the first feature. The method followed here is inspired by Reference [40]. A normalized window of *1-second* duration in the time series data is chosen to carry out the feature extraction.

First, a lowest contour line in the *1-second* data window circulating a local maximum point has been detected. Then the height of that point is measured and a ratio is prepared in terms of that contour line. Figure 9 shows the method of finding prominent local maxima from a normalized window of the principle component of one of the generators. For clarity of vision a *2-second* data window is selected for Figure 9. Once several prominent peaks are detected, the maximum available prominent peak is considered as a feature-factor for that data window and stored in the training data table.

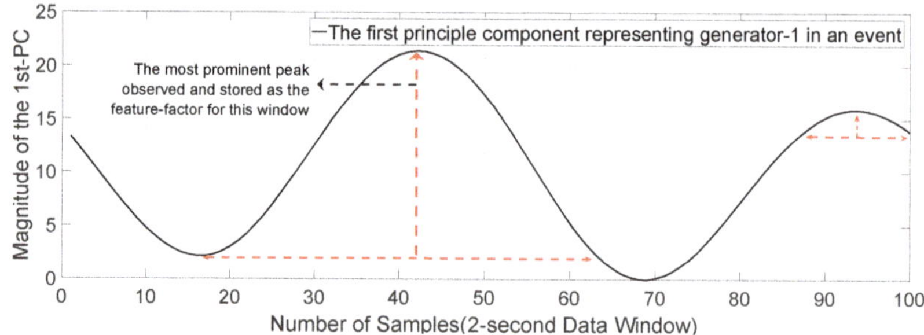

Figure 9. Peak prominence and width inside a predefined data window of the Terminal voltage of Generator-1.

The second feature is based on the available frequencies in the first principle component, observed in the *1-second* data window. This feature is extracted through a Discrete Fourier Transformation (DFT) technique; $x[n] = \sum_{n=0}^{N-1} \frac{1}{N}\tilde{X}[k]e^{-jk(2\pi/N)/n}, K = 0, 1, ...N-1$. Here, $\tilde{X}[k]$ is the amplitudes and $x[n]$ is the linear combination of the complex exponentials with that amplitude. Decision making based on frequency data has often been observed in the field of harmonics and power quality analysis [41]. In this study, a frequency spectrum of the first 30 Hz has been chosen as the featured attribute as shown in Figure 10 (with a small displacement in the X-axis for the purpose of visualization). As the sampling rate used in this study is 50 samples/seconds, according to the so called 'Nyquist-theorem' DFT can observe upto 30 Hz. Once the magnitudes of those frequencies are calculated, the magnitudes are summed up to be used as a feature-factor for that data window, $\sum_{i=1}^{30} M_i$. It is called a frequency factor.

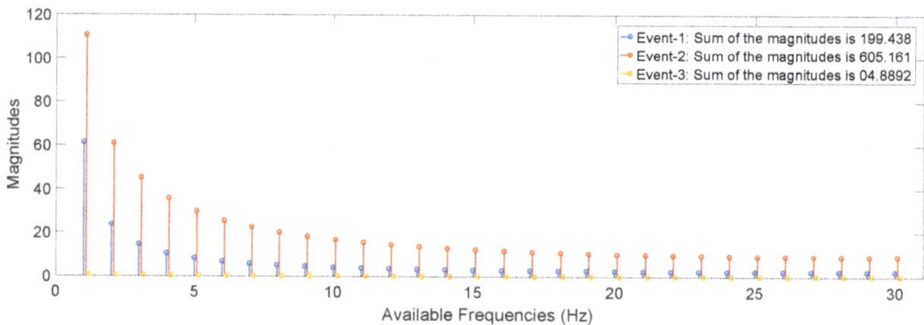

Figure 10. Available frequencies. The sum of all the available frequencies is presented as the frequency factor.

A sensitivity data, based on the voltage at fault bus and the active power generation from the subject generator is considered as the third feature [1].

In Figure 11, which is inspired by Liapunov's direct method, three sensitivity curves of the generators {G7, G1, G8} are shown in terms of the fault bus voltage at bus-16. It clearly shows the aperture of the curve varies depending on the impact of fault. The figure refers to the idea that, less sensitive relations have a larger aperture. To calculate the aperture, trapezoid method of numerical integration is used. The limit of each response curve is considered between two extreme points over the *X*-axis or the ΔV_{fbus} axis. Two extreme points develops a convex (Upper) and a concave (Lower) perimeters. Then the area under the curves are calculated. The overall area of the aperture is then assumed by subtracting the area under the concave curve from the convex; $Area = A_{convex} - A_{concave}$. This area of aperture is then used as the feature-factor that represents the data window.

$$\int_{V1}^{V2} f(x_{convex})dx \equiv \frac{V2 - V1}{2N} \sum_{n=1}^{N} (f(x_n) + f(x_{n+1})) \tag{7}$$

$$\int_{V1}^{V2} f(x_{concave})dx \equiv \frac{V2 - V1}{2N} \sum_{n=1}^{N} (f(x_n) + f(x_{n+1})). \tag{8}$$

Once the training data-table is prepared, it is used to train a multiclass classifier. Overall, the training process only considers data that ensures system stability. It means the classification algorithm is trained based on the concept of resiliency. The decisions leading towards post fault stability that returns to pre-contingent state are considered accurate decisions.

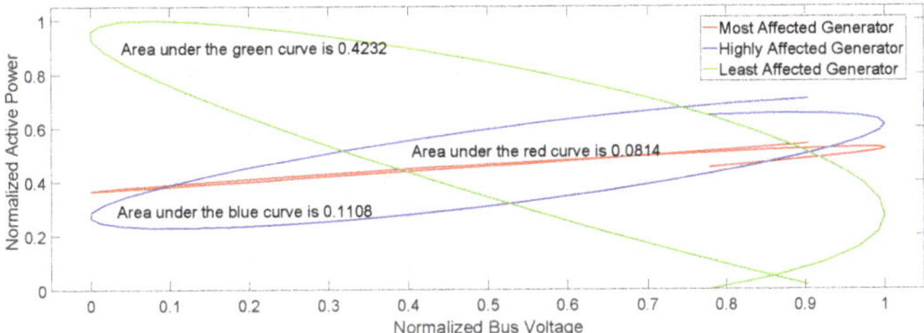

Figure 11. Sensitivity $\partial P_m / \partial V_{fbus}$ Curve.

3.5. Multiclass Classifier

The chosen Multiclass classifier is an ensemble of bagged decision trees trained using the stochastic parameters and the features. The bagged decision trees weigh and reweigh the predictor variables and their estimates. For example; a function estimate $\hat{g}_{ens} = \sum_{k=1}^{M} c_k \hat{g}_k(.)$ is obtained based on the $k - th$ reweighed data with the combined linear estimation co-efficient c_k. This method is deployed to eliminate classification error due to estimation variance and also statistical bias, specially in a stochastic scenario [32,42]. The ensemble method implemented in this study is prepared using one hundred fully grown trees by splitting the attribute data into one hundred training sets, D_1, D_2, ..., D_{100}. This approach helps to obtain an improved composite model. $M_i (1 \geq i \geq 100)$ classifiers vote by predicting a class and the ensemble selects the final class from those votes. The overall technique along with *Random Forest* and *Random Subspace* also includes *Bagging* and *Boosting* [36]. Depending on the number of controlled islands observed in each contingency, different values of the features are obtained. Three separate tables Tables 2–4 are shown below based on the number of islands. Each table shows the attributes *wind speed*, *Variable Demand*, *Snsvt* which is the feature-factor derived after sensitivity analysis, *Prmn* is the prominent maximam peak obtained from the principle component, *FF* is the frequency-factor derived from the DFT of the principle component. The tables are prepared from the data collected from generator-4. The classifier has a target attribute named *Decisions*. The decisions are combination of active power reference (0-1 P.U.) for the governor of the generator and the amount of temporary load shedding required in the Area-1. Both the information has been collected from the offline simulation after conducting the CC on each of the generators. n-number of decision combinations are observed and it is different for different generators, for example, for a two-area solution; in case of generator-1 $n = 4$ and for generators-4&7 $n = 7$. The control parameters are unique for each generator however the amount of load shedding required in each scenario is considered through a voting based on statistical mode [2]. To synchronize the voting process the load required to be shed is represented as segments of 100 MW, 200 MW, 300 MW up to 500 MW. An example of the decisions is shown in Figure 12. The upper figure shows the combination of two parameters observed in the decisions in a generator. The figure shows how the voting process and the active power reference are related in one generator under multiple scenarios. Both the estimations are carried out by the multiclass classifiers placed in each generating stations. The lower figure shows an instance of the voting process from the 10 synchronous machines in one scenario. As segment-4 got the maximum vote 4×100 MW load is temporarily shed from the Area-1.

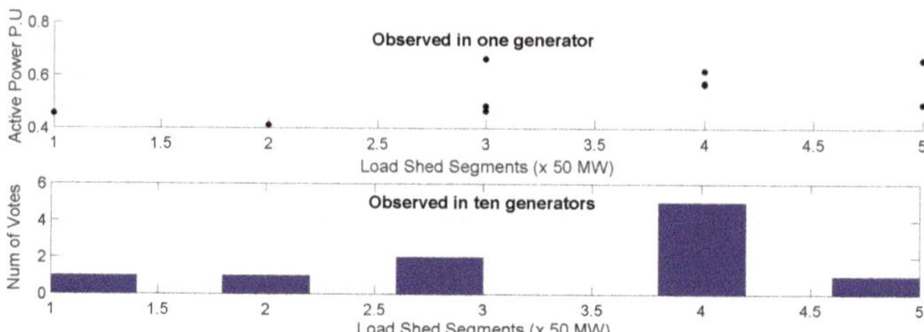

Figure 12. A set of example decisions: Observed from individual generator under different scenarios and an instance of statistical voting observed from all the generators under one scenario.

The decision trees, prepared from the above mentioned method, are shown in Figure 13 with a small example data set. The upper tree shows the decisions for selecting the amount of load shedding required in Area-1 . Data of the right tree is collected from all the synchronous generators and is used for the voting process for the purpose of load shedding, as shown in Figure 12. The left tree shows the active power reference set in that generator. The trees are prepared using generator-4 and time-line-1 data.

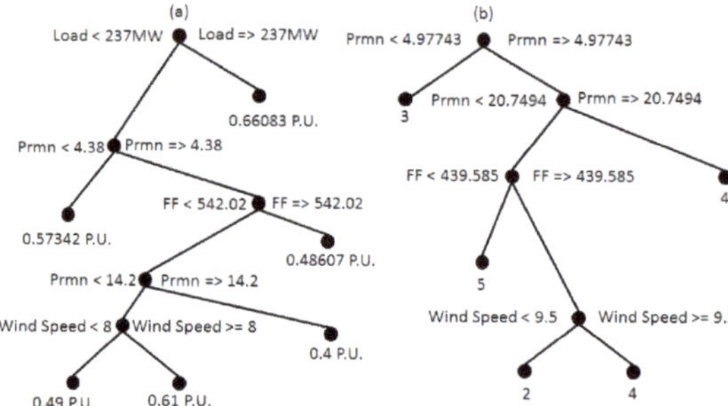

Figure 13. An instance of the proposed classification process in Generator-4. (**a**) The tree that selects active power reference using features (**b**) The tree that carries out priority voting.

The classification accuracy is measured against the system resiliency during low probability high disruptive events. If the classification algorithm can predict decisions that lead towards a stable pre-contingent state the algorithm is considered to be accurate.

Table 2. Distributed decision table for generator-4. Contingencies leading to a two area solution.

Decisions	Wind Speed (m/s)	Variable Demand (MW)	Snsvt (Sensitivity)	Prmn (Prominence)	FF (Frequency Factor)
1	12	35	0.1384	34.1	197.4
2	6	145	0.1333	0.659	34.48
3	8	405	0.1636	0.0566	7.5878
..	..	..	..	..	..
..	..	..	..	..	..
n	11	790	0.1424	0.593	25.2

Table 3. Distributed decision table for generator-4. Contingencies leading to a three area solution.

Decisions	Wind Speed (m/s)	Variable Demand (MW)	Snsvt (Fault-1)	Snsvt (Fault-2)	Prmn (Prominence)	FF (Frequency Factor)
1	8.2	662	0.8770	1.0195	0.4374	13.7012
2	10	372	0.9349	0.9910	0.7647	178.1212
..	..	..	..	..	..	..
..	..	..	..	..	..	..
n	11.2	408	0.8515	0.2779	0.7099	32.4

Table 4. Distributed decision table for generator-4. Contingencies are not followed by multiple area segmentation.

Decisions	Wind Speed (m/s)	Variable Demand (MW)	Snsvt (Sensitivity)	Prmn (Prominence)	FF (Frequency Factor)
1	9.9965	252	0.2192	0.0008	46.7454
2	10	546	0.2119	0.2541	46.8115
..	..	..	..	..	..
..	..	..	..	..	..
n	12	609	0.1547	0.1291	9.56

4. Results

Figure 14 shows three scenarios of the three proposed solutions under multiple contingencies and operational decisions as explained in the earlier sections. One interesting observation can be made that, in different stochastic scenarios the duration as well as the sequences of the timelines vary, specially for the scenarios where multiple faults lead towards a three area solution. The system resiliency is considered based on the post restoration terminal voltages. If the consumers are restored and the system voltage remains stable, the algorithm is considered to be improving resiliency.

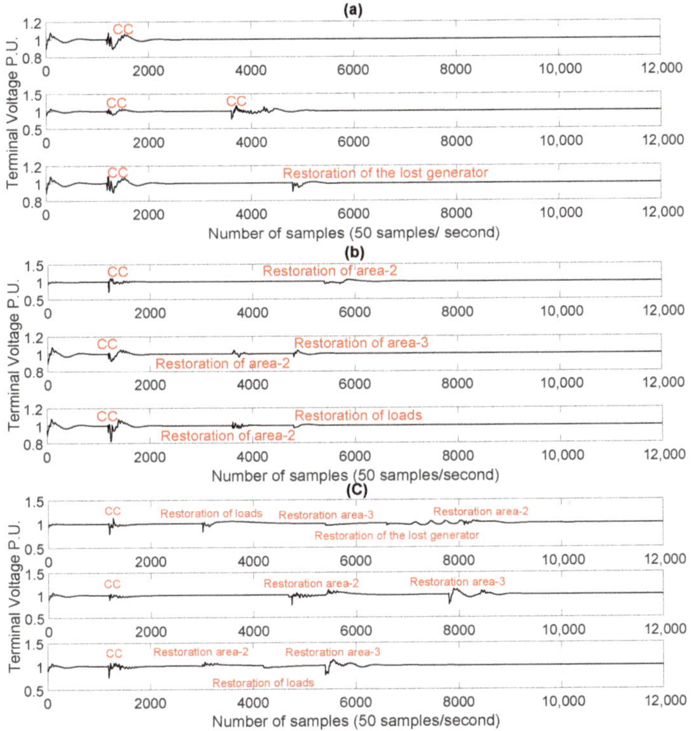

Figure 14. Three cases of grid restoration. (**a**) No segmentation solution, (**b**) Two-area solution, (**c**) Three-area solution.

Each of these time-lines has been assessed using Monte Carlo based simulation strategy by randomly generating wind power and demand. Based on these randomly selected stochastic scenarios, n_i decisions have been identified in this study, where n_i means n combination of decisions for the i-th generator. These decisions can maintain a post restoration stability. For example; for the two-area based solution, the Monte Carlo based simulation is carried out for a limited 1560 scenarios. Where, 1000 scenarios have been used for training the algorithm and rest of the 560 scenarios have been used for testing. Figure 15 shows a comparison of the average prediction accuracy between the proposed method with features as well as the prediction without features. Only the first principle component has been used as raw data where no feature has been extracted. If the system is completely restored the accuracy is considered '1' else '0'. The overall performance shows around 95% prediction accuracy with the features. With a limited 1000 scenarios, feature data has been proven to be quite helpful to reach such high degree of accuracy. The proposed accuracy is also a marker for symbolizing system resiliency. If the therefore, 95% accuracy also refers that in 95% cases the system remains resilient.

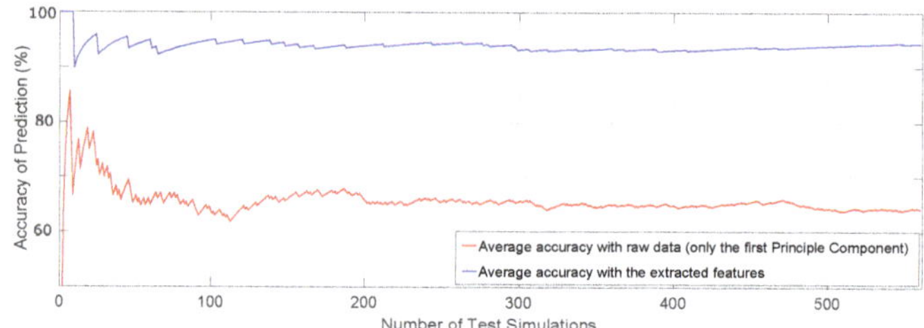

Figure 15. Prediction accuracy with the test data (two-area solution). The comparison shows a clear improvement with the proposed feature-base method.

The rest of the 5% cases where the algorithm misclassified the decisions, the system recovery could not be achieved. Once the areas are restored through the transmission line the overall system became unstable.

The proposed algorithm has also been compared with the similar methods proposed for the purpose of power system event detection in the previous literature [10,17]. Table 5, shows the comparison of accuracies in identifying 'timeline-1' under three different scenarios. The three scenarios are randomly selected among three-area solutions, two-area solutions and the no-segmentation solutions. Due to having additional information in the form of features, the proposed algorithm is demonstrating superiority. Figure 7 can be brought as a point of reference for explaining such performance. The overlapping magnitudes of the principle components, makes it difficult to trace out a linearized distinction between different events. The features introduces that linearity in data segmentation for the decision trees.

Table 5. Comparison in accuracies for *Timeline-1* in Percent %.

Generator	Proposed Method	Ensemble of Clusters and Decision Tree	ANN	PCA and Decision Tree
4	96.2	89.4	<40	83.71
5	96.4	96.4	<40	82.02
6	94.64	94.56	<40	84.2
7	95.08	93.33	<40	65.72
8	94.32	90.17	<40	67.23
9	95.48	92.8	<40	66.49
10	93.8	91.1	<40	89.7

The ramifications of voting for load shed has also been speculated. It has been observed that, if the required load shedding is not carried out the system could not be fully restored. Figure 16 shows a comparison between these cases. In the top figure a comparison is shown, where in one scenario the algorithm successfully identified the right decision and in the next the algorithm could not. In the lower half of the figure, three incidents are shown where the load shedding is not carried out based on the majority voting. In each of these cases the system has moved back towards instability.

In the Figures 14–16 the superiority of the proposed method can be observed. One critical observation regarding the scenarios having multiple faults is that, only the faults occurring simultaneously have been considered. In future studies different times of fault-occurrence can be analyzed.

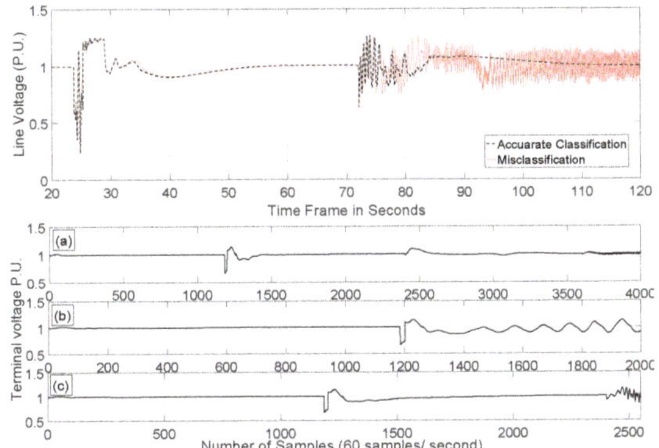

Figure 16. Ramifications of the 'Classification Error'. Top figure: Misclassification in active power referencing. Bottom figure: Misclassification in load-shedding; (**a**) Instability observed after grid restoration, (**b**) Instability observed after islanding, (**c**) Instability observed after the load has been restored.

5. Conclusions

The proposed method to predict suitable decisions for post fault restoration, has shown promising results. The method works with high accuracy despite limited number of training data has been provided. IEEE-39 bus test system, which is a large enough dynamic system, has been used to demonstrate the capabilities of the proposed method. Furthermore, the proposed algorithm is tested with high degree of stochastic influence in the network. This study is highly significant for a stand alone system where multi-machine dynamics can be observed. Based on the data analysis carried out in this study it can be concluded that, simplified features based on domain knowledge can be highly effective for the purpose of service restoration.

One critical observation is that, with an increased number of faults, the system needs to be divided into more than two areas. Under those cases the proposed algorithm heavily relies on the number of training data set provided. It is because, the optimization process in this case has to consider the sequence of operations. Thus, in case of new data sets with multiple fault scenarios, the accuracy of the algorithm decreases compare to those of the single contingency scenarios. In the author's future experiments, the role of a centralized control scheme to address the problem of sequence control would be analyzed and compared with the proposed distributed method. Overall it can be stated that, the proposed method is showing significant promise in the filed of grid restoration techniques.

The algorithm is tested online basis. However, it is not tested on a real-time platform. Therefore, the time consumption for the decision making process varies case by case. In future studies it would be more comprehensive to apply the method on a real-time platform to address time-consumption by the system and its impact on a real-time decision-making process.

Author Contributions: M.A.K. proposed the approach, contributed in literature review, modelling, simulation, test study and in the manuscript write-up. J.C. and T.-T.L. contributed in critical review and in proofreading of the manuscript. All authors have read and agreed to the published version of the manuscript.

Funding: This research received no external funding.

Conflicts of Interest: The authors declare no conflict of interest.

References

1. Zhang, J.; Zheng, X.; Wang, Z.; Guan, L.; Chung, C.Y. Power System Sensitivity Identification 2014;Inherent System Properties and Data Quality. *IEEE Trans. Power Syst.* **2017**, *32*, 2756–2766. [CrossRef]
2. Al Karim, M.; Currie, J.; Lie, T.T. A machine learning based optimized energy dispatching scheme for restoring a hybrid microgrid. *Electr. Power Syst. Res.* **2018**, *155*, 206–215. [CrossRef]
3. Shahnia, F.; Chandrasena, R.P.S.; Rajakaruna, S.; Ghosh, A. Primary control level of parallel distributed energy resources converters in system of multiple interconnected autonomous microgrids within self-healing networks. *IET Gener. Transm. Distrib.* **2014**, *8*, 203–222. [CrossRef]
4. Konstantelos, I.; Jamgotchian, G.; Tindemans, S.H.; Duchesne, P.; Cole, S.; Merckx, C.; Strbac, G.; Panciatici, P. Implementation of a Massively Parallel Dynamic Security Assessment Platform for Large-Scale Grids. *IEEE Trans. Smart Grid* **2017**, *8*, 1417–1426. [CrossRef]
5. Zidan, A.; Khairalla, M.; Abdrabou, A.M.; Khalifa, T.; Shaban, K.; Abdrabou, A.; Shatshat, R.E.; Gaouda, A.M. Fault Detection, Isolation, and Service Restoration in Distribution Systems: State-of-the-Art and Future Trends. *IEEE Trans. Smart Grid* **2017**, *PP*, 1–16. [CrossRef]
6. Zhou, Y.; Arghandeh, R.; Konstantakopoulos, I.; Abdullah, S.; von Meier, A.; Spanos, C.J. Abnormal event detection with high resolution micro-PMU data. In Proceedings of the 2016 Power Systems Computation Conference (PSCC), Genoa, Italy, 20–24 June 2016; pp. 1–7. [CrossRef]
7. Jin, S.; Huang, Z.; Diao, R.; Wu, D.; Chen, Y. Comparative Implementation of High Performance Computing for Power System Dynamic Simulations. *IEEE Trans. Smart Grid* **2017**, *8*, 1387–1395. [CrossRef]
8. Khaitan, S.K. A survey of high-performance computing approaches in power systems. In Proceedings of the 2016 IEEE Power and Energy Society General Meeting (PESGM), Boston, MA, USA, 17–21 July 2016; pp. 1–5. [CrossRef]
9. Vasconcelos, M.H.; Carvalho, L.M.; Meirinhos, J.; Omont, N.; Gambier-Morel, P.; Jamgotchian, G.; Cirio, D.; Ciapessoni, E.; Pitto, A.; Konstantelos, I.; et al. Online security assessment with load and renewable generation uncertainty: The iTesla project approach. In Proceedings of the 2016 International Conference on Probabilistic Methods Applied to Power Systems (PMAPS), Beijing, China, 16–20 October 2016; pp. 1–8. [CrossRef]
10. Chen, Y.; Xie, L.; Kumar, P.R. Power system event classification via dimensionality reduction of synchrophasor data. In Proceedings of the 2014 IEEE 8th Sensor Array and Multichannel Signal Processing Workshop (SAM), A Coruna, Spain, 22–25 June 2014; pp. 57–60. [CrossRef]
11. Mohammadi, M.; Gharehpetian, G.B. Application of core vector machines for on-line voltage security assessment using a decisiontree-based feature selection algorithm. *IET Gener. Transm. Distrib.* **2009**, *3*, 701–712. [CrossRef]
12. Jain, A.; Zongker, D. Feature selection: Evaluation, application, and small sample performance. *IEEE Trans. Pattern Anal. Mach. Intell.* **1997**, *19*, 153–158. [CrossRef]
13. Niazi, K.; Arora, C.; Surana, S. Power system security evaluation using ANN: Feature selection using divergence. *Electr. Power Syst. Res.* **2004**, *69*, 161–167. [CrossRef]
14. Geeganage, J.; Annakkage, U.D.; Weekes, T.; Archer, B.A. Application of Energy-Based Power System Features for Dynamic Security Assessment. *IEEE Trans. Power Syst.* **2015**, *30*, 1957–1965. [CrossRef]
15. Abedinia, O.; Amjady, N.; Zareipour, H. A New Feature Selection Technique for Load and Price Forecast of Electrical Power Systems. *IEEE Trans. Power Syst.* **2017**, *32*, 62–74. [CrossRef]
16. dos Santos, A.; de Barros, M.T.C. Sensitivity of voltage sags to network failure rate improvement. In Proceedings of the 2016 Power Systems Computation Conference (PSCC), Genoa, Italy, 20–24 June 2016; pp. 1–7. [CrossRef]
17. Papadopoulos, P.N.; Guo, T.; Milanović, J.V. Probabilistic Framework for Online Identification of Dynamic Behavior of Power Systems With Renewable Generation. *IEEE Trans. Power Syst.* **2018**, *33*, 45–54. [CrossRef]

18. Athay, T.; Podmore, R.; Virmani, S. A Practical Method for the Direct Analysis of Transient Stability. *IEEE Trans. Power Appar. Syst.* **1979**, *PAS-98*, 573–584. [CrossRef]
19. Bidram, A.; Davoudi, A. Hierarchical structure of microgrids control system. *Smart Grid IEEE Trans.* **2012**, *3*, 1963–1976. [CrossRef]
20. Arunagirinathan, P.; Abdelsalam, H.A.; Venayagamoorthy, G.K. Remote power system stabilizer tuning using synchrophasor data. In Proceedings of the 2014 IEEE Symposium on Computational Intelligence Applications in Smart Grid (CIASG), Orlando, FL, USA, 9–12 December 2014; pp. 1–7. [CrossRef]
21. Grondin, R.; Kamwa, I.; Trudel, G.; Gerin-Lajoie, L.; Taborda, J. Modeling and closed-loop validation of a new PSS concept, the multi-band PSS. In Proceedings of the 2003 IEEE Power Engineering Society General Meeting (IEEE Cat. No.03CH37491), Toronto, ON, Canada, 13–17 July 2003; Volume 3, p. 1809. [CrossRef]
22. Lee, D. Ieee recommended practice for excitation system models for power system stability studies (ieee std 421.5-1992). *Energy Dev. Power Gener. Comm. Power Eng. Soc.* **1992**, *95*, 96.
23. Grondin, R.; Kamwa, I.; Soulieres, L.; Potvin, J.; Champagne, R. An approach to PSS design for transient stability improvement through supplementary damping of the common low-frequency. *IEEE Trans. Power Syst.* **1993**, *8*, 954–963. [CrossRef]
24. Liu, D.; Antsaklis, P.J. *Stability and Control of Dynamical Systems with Applications: A Tribute to Anthony N. Michel*; Springer Science & Business Media: Berlin, Germany, 2012.
25. Teymouri, F.; Amraee, T.; Saberi, H.; Capitanescu, F. Towards Controlled Islanding for Enhancing Power Grid Resilience Considering Frequency Stability Constraints. *IEEE Trans. Smart Grid* **2017**, *10*, 1735–1746. [CrossRef]
26. Rabiee, A.; Soroudi, A.; Keane, A. Risk-Averse Preventive Voltage Control of AC/DC Power Systems Including Wind Power Generation. *IEEE Trans. Sustain. Energy* **2015**, *6*, 1494–1505. [CrossRef]
27. Babazadeh, M.; Karimi, H. Robust decentralized control for islanded operation of a microgrid. In Proceedings of the 2011 IEEE Power and Energy Society General Meeting, Detroit, MI, USA, 24–28 July 2011; pp. 1–8. [CrossRef]
28. Vittal, E.; O'Malley, M.; Keane, A. Rotor angle stability with high penetrations of wind generation. *IEEE Trans. Power Syst.* **2012**, *27*, 353–362. [CrossRef]
29. Hijazi, H.L.; Mak, T.W.; Van Hentenryck, P. Power System Restoration With Transient Stability. In Proceedings of the AAAI, Austin, TX, USA, 25–30 January 2015; pp. 658–664.
30. Wang, Z.; Makram, E.B.; Venayagamoorthy, G.K.; Ji, C. Dynamic estimation of rotor angle deviation of a generator in multi-machine power systems. *Electr. Power Syst. Res.* **2013**, *97*, 1–9. [CrossRef]
31. Rabiee, A.; Soroudi, A.; Mohammadi-ivatloo, B.; Parniani, M. Corrective Voltage Control Scheme Considering Demand Response and Stochastic Wind Power. *IEEE Trans. Power Syst.* **2014**, *29*, 2965–2973. [CrossRef]
32. Genc, I.; Diao, R.; Vittal, V.; Kolluri, S.; Mandal, S. Decision Tree-Based Preventive and Corrective Control Applications for Dynamic Security Enhancement in Power Systems. *IEEE Trans. Power Syst.* **2010**, *25*, 1611–1619. [CrossRef]
33. Karim, M.A.; Currie, J.; Lie, T.T. A distributed machine learning approach for the secondary voltage control of an Islanded micro-grid. In Proceedings of the 2016 IEEE Innovative Smart Grid Technologies—Asia (ISGT-Asia), Melbourne, Australia, 28 November–1 December 2016; pp. 611–616. [CrossRef]
34. Guo, T.; Milanović, J.V. Online Identification of Power System Dynamic Signature Using PMU Measurements and Data Mining. *IEEE Trans. Power Syst.* **2016**, *31*, 1760–1768. [CrossRef]
35. Nguyen, T.T.; Nguyen, V.L. Power System Security Restoration by Secondary Control. In Proceedings of the 2007 IEEE Power Engineering Society General Meeting, Tampa, FL, USA, 24–28 June 2007; pp. 1–8. [CrossRef]
36. Zhang, C.; Li, Y.; Yu, Z.; Tian, F. Feature selection of power system transient stability assessment based on random forest and recursive feature elimination. In Proceedings of the 2016 IEEE PES Asia-Pacific Power and Energy Engineering Conference (APPEEC), Xi'an, China, 25–28 October 2016; pp. 1264–1268. [CrossRef]
37. Jurado, S.; Nebot, À.; Mugica, F.; Avellana, N. Hybrid methodologies for electricity load forecasting: Entropy-based feature selection with machine learning and soft computing techniques. *Energy* **2015**, *86*, 276–291. [CrossRef]
38. Honkala, I.S.; Laihonen, T.K. The probability of undetected error can have several local maxima. *IEEE Trans. Inf. Theory* **1999**, *45*, 2537–2539. [CrossRef]

39. Abel, E.W.; Arkidis, N.S.; Forster, A. Inter-scale local maxima-a new technique for wavelet analysis of EMG signals. In Proceedings of the 20th Annual International Conference of the IEEE Engineering in Medicine and Biology Society. Vol.20 Biomedical Engineering Towards the Year 2000 and Beyond (Cat. No.98CH36286), Hong Kong, China, 1 November 1998; Volume 3, pp. 1471–1473. [CrossRef]
40. Griffié, J.; Boelen, L.; Burn, G.; Cope, A.P.; Owen, D.M. Topographic prominence as a method for cluster identification in single-molecule localisation data. *J. Biophotonics* **2015**, *8*, 925–934. [CrossRef]
41. Hu, Z.; Xiong, M.; Shang, H.; Deng, A. Anti-Interference Measurement Methods of the Coupled Transmission-Line Capacitance Parameters Based on the Harmonic Components. *IEEE Trans. Power Deliv.* **2016**, *31*, 2464–2472. [CrossRef]
42. Bühlmann, P. Bagging, boosting and ensemble methods. In *Handbook of Computational Statistics*; Springer: Berlin/Heidelberg, Germany, 2012; pp. 985–1022.

MDPI
St. Alban-Anlage 66
4052 Basel
Switzerland
Tel. +41 61 683 77 34
Fax +41 61 302 89 18
www.mdpi.com

Energies Editorial Office
E-mail: energies@mdpi.com
www.mdpi.com/journal/energies

www.ingramcontent.com/pod-product-compliance
Lightning Source LLC
LaVergne TN
LVHW070704170726
843515LV00004B/479

* 9 7 8 3 0 3 6 5 3 6 5 3 8 *